# Being human

# Being human

## HISTORICAL KNOWLEDGE AND
## THE CREATION OF HUMAN NATURE

Roger Smith

Columbia University Press — New York

Columbia University Press
*Publishers Since 1893*
New York

Copyright © 2007 Roger Smith

First published in the United Kingdom by Manchester University Press
All rights reserved

Library of Congress Cataloging-in-Publication Data

Smith, Roger, 1945–
Being human : historical knowledge and the creation of
human nature / Roger Smith.
p. cm.
Includes bibliographical references and index.
ISBN 978–0–231–14166–6 (cloth : alk. paper) —
ISBN 978–0–231–51290–9 (ebook)
1. Philosophical anthropology. I. Title.

BD450.S5553 2007
128—dc22

A complete CIP record is available from the Library of Congress

# Contents

Preface      *page* vi

Introduction      1

1   Being human      16

2   Reflexive knowledge      62

3   Relations of the natural and human sciences      93

4   Precedents for the human sciences      122

5   Historical knowledge      173

6   Values and knowledge      197

Epilogue: on human self-creation      243

Bibliography      260

Index      281

# Preface

How do we represent our own being to ourselves? How do we address the question 'What is human?' About this, everyone may feel they have something to say.

The purpose of my approach to the question is to link different notions of being human with historical knowledge and, in the process, to clarify and compare different views of being human in the natural sciences, the social sciences and the humanities. For these ends, it is necessary to argue on a broad front. The book therefore takes a large canvas on which to picture western thought on human nature. It is not a specialised study in any one field; all the same, broad perspectives have their own discipline.

I seek to navigate the shoals of human self-knowledge as a historian. Whatever the final destination, these arguments, as it happens, began with an attempt to explain the scholarly activity called 'the history of the human sciences'. This opened, inevitably, into a much large domain. Since I have been associated with a journal, *History of the Human Sciences*, and wrote a narrative history of the human sciences, people have, on a number of occasions, asked me to summarise the nature and purposes of the field. But I remained dissatisfied. Now I have tried to do justice to the arguments. This has taken matters far beyond their starting point and into complex philosophical questions about human self-understanding. These questions I approach as a historian seeking to know why history matters.

For economy and readability, I try to sustain the argumentative form of an 'essay'. If the reader is interested to know more history, or more historical sources, he or she is invited to turn to my earlier book. There are ample more specialised studies in particular philosophical topics.

A book such as this crosses boundaries and touches on many interests. I am accordingly indebted to many people, institutions and discussions, first and foremost, over many years, to Lancaster University and especially the intellectual history seminar in the Department of History. I also gratefully acknowledge the British Society for the History of Science symposium at the Science Museum, London, on 'The big picture', in 1993; the Institutionen för Idé- och Lärdomshistoria, Göteborgs Universitet, which was home to me in 1993–94; a *colloque* of the Société Française pour l'Histoire des Sciences de l'Homme in Paris, 1996; a conference on 'Writing the history of the human sciences', Centre for Cultural History at the University of Aberdeen, in 1997, organised by George Rousseau; an invitation in 1997 to Cheiron: The International Society for the History of the Behavioral and Social Sciences, at the University of Virginia in Richmond; a fellowship at the Dibner Institute for the History of Science and Technology, at MIT in Cambridge, Massachusetts, 1997–98; the Centre de Recherche en Histoire des Sciences et des Techniques, Cité des Sciences et de l'Industrie, Paris, 1998–99; colloquia at the Boston University Center for the Philosophy and History of Science on 'The boundary of the human sciences' in 1998 and on 'Reflexivity redux' in 2003, organised by Alfred I. Tauber; a *colloque* on 'Unité et globalité de l'homme: des humanités aux sciences humaines', organised at Université Paris VII by Jacqueline Carroy, Claude Debru and Marie-Louise Pelus-Kaplan, in 2002; a workshop, 'Versions of expertise: reconstructing subjects and subjectivity in the human sciences', of the European Union INTAS project 671–30631, at Universiteit Groningen, in 2002; an East–West conference on 'Moral philosophy in pluralistic cross-cultural context', Institute of Philosophy of the Russian Academy of Sciences, Moscow, organised by Marietta Stepanyants, in 2002; the annual conference of the European Society for the History of the Human Sciences, meeting in the Universitat Autónoma de Barcelona, also in 2002. In 2004, I was much encouraged to clarify my thoughts in the Institut Universitaire Romand d'Histoire de la Médecine et de la Santé, in Lausanne, directed by Vincent Barras. Jan Golinski and the departments of History and of Psychology, University of New Hampshire, arranged for me to argue 'why history matters' in 2005. The British Library has been of inestimable value to me, as to so many other people.

I warmly thank editors of journals and books who have encouraged me to approach 'the big picture', including Claude Blanckaert, John C. Burnham, Jim Good, Sheila Jasanoff, the late Roy Porter, Asya Syrodeeva and the late Irving Velody. Many people have helped with information or tried to put me straight. I thank a generous Germanist who read my text for the Press. Thanks to Constance Blackwell, Juliet and Brian Braithwaite, Roger Cooter, Rhodri Hayward, Graham Richards and Robert M. Young, a friend over many years, in London.

The book owes a great deal to the thoughtful support and criticism of Jan Golinski, Irina Sirotkina and Alfred I. Tauber.

# Introduction

What is human nature? The words of the question in an English-language scientific culture almost presuppose an answer in terms of biology. Yet we cannot properly answer the question without first addressing complexities in human self-knowledge. Human nature is not some 'thing' waiting discovery but active in understanding itself.

This book seeks to think about human nature, or, as I prefer to say, being human, in relation to historical knowledge. In common English-language usage 'science' means 'natural science'. The fact is, however, that there is an alternative tradition. In continental European usage, 'science' is a body of knowledge thought to be rationally grounded and hence true; it is therefore common to refer to the disciplines of the study of literature and art, or of history and geography, or of the mind and conduct and of the social world generally, and even of theology, as 'sciences'. Some, but by no means all, contemporary English-language writers follow this tradition when they refer to 'the human sciences'. What I do is look again at the continental European practice and, especially, at the notion that history is both a science and the key to human self-understanding, that is, to knowledge of being human.

Seeking self-knowledge, we divide ourselves into two and become both subject and object. 'We ourselves are not only the beings, that reason, but also one of the objects, concerning which we reason.'[1] To know oneself is to launch oneself

---

1 Hume, *Treatise of Human Nature*, p. xix. (Throughout, I adopt the original printed form of quoted passages, unless otherwise stated.)

into the air, so to speak, while yet staying on the ground. Gothic writers in the early nineteenth century found a powerful image for this in the *Doppelgänger*, the mysterious figure one encounters who appears to be oneself. 'But when I reach the door of my house and have got the keys in my hand, I – yes, I myself – am already standing before it and gazing wildly at myself with the same big black eyes as are to be found in my own head.'[2] In these circumstances, where can a person stand in order to get an objective view of human nature? The question is an existential one – what sort of being is it that can know itself? – and it is also a cognitive one. As a cognitive question, it is at the centre of the human sciences, the sciences about what people are and do. Any term we care to take to describe people in general – human beings, people, *Homo sapiens*, the human race, human nature, the naked ape, man, children of God – presupposes one point of view about being human. The language, with its rich connotations, contains a view about what we are before we begin any enquiry. This subject 'we' is an exemplary case in point: who are 'we' here?

The French philosopher and self-appointed head of a new church, Auguste Comte, was particularly scathing about the claims of psychologists of his day, the 1820s and 1830s, to observe the workings of their own minds: 'A thinking individual cannot divide himself into two, one half reasoning, and the other watching reason . . . Our posterity will doubtless one day see these pretensions transferred to the comic stage.' Introspection cannot be a source of knowledge. Instead, he argued, scientific knowledge of individual people must come from 'direct observation' of their physiology and their place in the social world.[3] What we can know, he claimed, is restricted to sensory perception of the world around us. Many scientists favour this position: if we want to know people, we should study their physical state and visible doings, just like we study anything else in nature. Yet, many other people, scientists as well as philosophers, have thought Comte completely wrong, principally because his way of thinking cannot make sense of

2 Hoffmann, 'The choosing of the bride', p. 370.
3 *The Essential Comte*, pp. 33, 56.

his own act of making statements about the world and knowledge of it. In spite of his appeal to 'direct observation', he did not escape being the subject as well as the object of observation. Moreover, who, or what, is active in observation – a biological brain, Comte the historical person, a self, the unconscious mind, language? Comte's attempt to sidestep the question, though greatly refined in the twentieth-century psychological and social sciences, has not stood up.

There is an alternative approach through historical understanding. This involves knowing what people have said and believed about being human. It also involves going beyond empirical description in order to claim that a right understanding of historical knowledge is essential to knowledge of being human. This is the thesis the book argues for. It is a thesis opposing common philosophical, biological, religious or ethical claims to define what being human 'really' means. It maintains that there is no position outside the historical forms which human life takes for an absolutely objective or eternally valid view.

David Hume, writing in the mid-eighteenth century, voiced what has remained a more popular view. Looking about him, he concluded that human nature has in its essentials remained the same. People think, feel and act in set ways. We can, he thought, use social and historical observation to reveal the constants of human nature and thereby create what he and his contemporaries called 'the science of man'.[4] 'It is universally acknowledged that there is a great uniformity among the actions of men, in all nations and ages, and that human nature remains still the same, in its principles and operations ... Mankind are so much the same, in all times and places, that history informs us of nothing new or strange in this particular. Its chief use is only to discover the constant and universal principles of human nature.'[5] Yet, for every statement identifying a seamless subject, then capitalised as 'Man', there was another statement drawing distinctions between people and describing their differences. Pascal, a century earlier, wrote:

4 See Fox, Porter and Wokler (eds), *Inventing Human Science.*
5 Hume, *Enquiries Concerning the Human Understanding*, p. 83.

'All is one, all is diversity. How many natures lie in human nature!'[6] Descriptions of difference separated people by individual character and occupation, as well as by their membership of groups like women, savages, gentlemen and peasants. This pattern, if not exactly the same categories with the same degree of self-assurance, persists. The sciences of human nature show a centripetal tendency, leading to the identification of a central core, and a centrifugal tendency, leading to the identification of difference. Then and now, observers like Hume have supposed, it is possible to find definite empirical answers, through sciences like psychology and anthropology, to questions about what causes similarity and what diversity.

Nevertheless, Hume was in part wrong. Firstly, he did not see how his own position within a rapidly developing commercial society affected the way he thought about human nature and what he thought common to human motives. This point applies to later theorists of human nature who take a norm for themselves to be a norm for all. Secondly, Hume did not consider that the most basic categories of human self-description (such as 'human' itself) have a history. But they do, and this is the basis of the claim, which I argue for, that the history of knowledge of being human is knowledge of being human, and the history of the human sciences is knowledge in the human sciences. Thirdly, Hume did not see that there is circularity in the way he, and everyone else, identified similarity and difference: people bring a prior view of what sort of being a human is to bear on their descriptions of people. There are no neutral descriptions. We need a historical and philosophical perspective in order to bring into the open the origins and implications of our self-descriptions.

It makes a difference which presumptions we have. If you think that to be a person is to be a moral agent, then it is always open to judge the rightness of what a person does. We may choose not to do so, and we may certainly prefer not to praise or blame, but the principle remains. Conversely, if we treat someone as not an agent, we do not assign her or him the same kind of dignity. If we define being a person as having a certain kind of natural, biological nature, something given

6 Pascal, *Pensées*, p. 32 (§ 129).

by the evolutionary past, then it does not seem possible to call an action caused by human nature right or wrong. This makes a great deal of difference emotionally and to the sense of self. A person who views her actions as the result of a common nature is not the same kind of person as someone who sees her actions as moral choices. People live differently as a result of differences in such beliefs. The issues are not esoteric. If you think jealousy is a matter of genes, your love affair will take a different course from that of someone for whom jealousy is a crisis in interpreting love.

Understanding and doing something about such differences of view requires stepping back from the issues immediately under debate in order to see why people conduct the debate in the terms they do. It requires deeper, mutual self-understanding. This does not come only from debating the supposed facts, but from looking into how we decide what facts are and how we shape the stories which arrange facts in a meaningful way. This book, therefore, does not state what human nature 'really' is. The philosophical questions at issue are far too complex for such statements. The question 'What is human?' does not have, and could not possibly have, one answer. There are different kinds of answer for different purposes, and there are good reasons for different conceptions of what it is to be human.

It is worth pausing to wonder what is being asked for from a theory of human nature. Given that the subject who asks the question is human too, and anything he or she asserts is a human statement, it looks as if the desired theory might be a theory of everything and, thereby, a theory of nothing. It is tempting to borrow Hegel's contemptuous words, written about the over-simple dualistic philosophies of nature of his day, to oppose ready formulas about human nature: 'The instrument of this monotonous formalism is no more difficult to handle than a painter's palette having only two colours . . . It would be hard to decide which is greater in all this, the casual ease with which everything in heaven and on earth and under the earth is coated with this broth of colour, or the conceit regarding the excellence of this universal recipe.'[7] Rather than laying

7 Hegel, *Phenomenology of Spirit*, pp. 30–1 (§ 51).

down formulas, perhaps we may look for understanding from the history of what we say about what people are.

It is a matter of report – hardly one from which there could be much dissension – that there is no contemporary unified human science, no unified body of knowledge about the human subject. A host of different sciences exist, and each claims knowledge and expertise for an area of human activity: anthropology, geography, economics, linguistics, neuroscience, psychoanalysis, political science and so on. Of course, there are scientists who regret this and hope unity will come, and there are even scientists, like the evolutionary psychologist Steven Pinker, who think they already possess the science which will create the unity. But on the ground, diversity is the reality; there are claims to human self-understanding right across the divisions known in the English-speaking world as the natural sciences, the social sciences and the humanities.

The view that there is a unified science, biology, which (in principle, if not yet in fact) accounts for everything at the base of human existence, presupposes that there is and can be only one possible foundation for knowledge of people, and it takes it as self-evident, in a scientific age, that this foundation is natural science. There is, however, a straightforward riposte to an exclusively biological approach. It is Roger Trigg's: 'Man is a biological species, but any attempt to assert that he is no more than that, with characteristics no different in kind from those of other such species, is not only mistaken. It is self-refuting, since no other species could formulate such a theory.'[8] I shall take the view that the evolutionary approach to human nature is one of a number of ways of thought people have constructed in the reflexive process of self-knowing. It is necessary to argue against reduction in all forms, that is, against claims that human nature is really, or merely, biological, or really economic or really moral. The subject, ourselves, is multi-dimensional in the largest sense, and we need knowledge that expresses this. All claims that one way of thought is 'basic' create abstractions – in A. N. Whitehead's excellent

8  Trigg, *Shaping of Man*, p. 174. Trigg, *Ideas of Human Nature*, introduces the philosophical argument that the nature of reasoning itself requires belief in a shared human identity.

phrase, they are examples of 'the fallacy of misplaced concreteness'.[9] We need to put ways of thought, evolutionary biology included, back into the historical context in which the particular human dimension they describe emerged as an object of knowledge. If we do not understand these contexts, we do not understand ourselves.

The word 'human' denotes something coming into existence in historical processes. No one would deny that the language used to describe people has changed. But the argument here makes the much stronger, and perhaps initially counterintuitive, claim that the subject matter which language describes has changed along with the language. To explain this involves coming to grips with some complex philosophical matters. It is a claim that many biologists, therapists and people with religious faith are likely to oppose. They want to assert that there is a real, lasting content to human nature, given by evolution, by the unconscious or by the creative act of a god. I do not assert the historical self-forming of being human as a way to deny or exclude such ways of thought. It is, in fact, part of the argument for the importance of historical knowledge that this knowledge explains why there are, legitimately and rationally, different ways of thought. All the same, I will assert as strongly as possible that theories, which claim exclusive access to knowledge of what is human, ignore history and refuse to examine their own historical contingency, are mistaken. This book's purpose is to put the historical point of view about the human subject in its rightful place. It is hard not to think that the current wave of enthusiasm for evolutionary and neuroscientific views of human nature, influenced by writers like Richard Dawkins and Daniel Dennett, has made many people blind to this. At the same time, the canonical status that certain writers, like Jacques Derrida or Michel Foucault, but only certain writers, have in what scholars in the social sciences and humanities call 'theory' is at the expense of knowledge of a host of other resources. In particular, the assumption that recent writing makes everything that has gone before uninteresting seems profoundly wrong.

9 Whitehead, *Science and the Modern World*, p. 64.

I interpret the answers people give to the question 'What is man?' – this was Kant's question – as a matter for historical knowledge. People are knowledge-making subjects who make themselves as they make knowledge. This is perhaps not an easy argument either to make or to grasp. All the same, it is an argument with a rich history, at least back to Giambattista Vico. One version appeared in Marx's writings: 'Man makes his life activity itself an object of his will and consciousness . . . Conscious life activity directly distinguishes man from animal life activity.'[10] People have made what they are over historical time: this is the key to being human, not knowledge of some unaltered core.

Several themes support the larger argument. For the first there is the term 'reflexivity'. This word characterises the nature and consequences of people being both subject and object of knowledge. Discussion of the rational conditions of knowledge (epistemology) shows that there is no body of knowledge free of presuppositions which the body of knowledge itself cannot validate. Put in perhaps over-simple terms, all knowledge involves an element of faith. There are always presuppositions open to critical analysis, and the critical analysis will itself contain unfounded presuppositions, and so on. One implication of this, since people are subject and object of knowledge, is that knowledge about human beings changes what people are. These arguments are central to the book's epilogue, which tries to make sense of the idea of human self-creation. This is a liberating prospect. Whatever the constraints that each and every person experiences, there is a potential for reflection, and that reflective act is in itself already a kind of change in a person, the basic element of a self-creation. Obviously, there are many circumstances in which, as a matter of fact, the possibility of such reflective activity is destroyed. But that does not render belief in its potential invalid.

The second theme is that historical work is necessary and not only valuable. Many people, not only historians, take this for granted. But many others, whether they are politicians allocating financial resources or natural scientists looking

10 Marx, 'Economic and philosophical manuscripts', p. 328.

forward to the next breakthrough, treat history as a luxury or even an indulgence. They echo Henry Ford's statement that history is more or less bunk. While not normally so rude, natural and social scientists often assume that empirical knowledge about what they can directly observe has some kind of natural priority over knowledge about the past. Historians, it would appear, often do not know how best to defend themselves. The serious response is that historians establish knowledge of a kind which other disciplines do not establish, and this knowledge includes knowledge of when, where and how scientists establish the kind of knowledge they do. If scientists, like other people, are to understand themselves, they too must have historical knowledge. I develop this thought into the general claim that without historical knowledge of the beliefs held about the nature of being human we are ignorant of what it is to be human. This gives content to R. G. Collingwood's aphorism: 'History is thus the self-knowledge of the living mind.'[11]

A third theme fosters a critical attitude to how people use the notion of human nature. Ordinary speech uses the term all the time to denote what is thought to be universal, or essential, to human beings, as in saying that it is human nature to be selfish. In Judeo-Christian context, the term summons up a picture of common descent from the Fall of Adam and Eve. In modern biological context, the term summons up a picture of a shared physical inheritance, leading, for example, to aggression, and there is a flourishing literature purporting to state what this inheritance actually is.[12] Given the complexity of human beings trying to understand themselves, it is perhaps no wonder that a way of thought, which appears to overcome difficulties with empirical facts, is attractive. This way of thought is heir to Victorian scientific naturalism, for which Darwin was the figurehead, treating people as knowable in the same manner as any other object in nature. Human nature,

---

11 Collingwood, *Idea of History*, p. 202.
12 Two influential programmatic statements are Wilson, *Sociobiology*, chapter 27, and Tooby and Cosmides, 'Psychological foundations of culture'. For evaluations see Kuper, *The Chosen Primate*, and Malik, *Man, Beast and Zombie*.

on this view, is 'out there' waiting to be uncovered. This way of seeing things necessarily changes, however, if we start to think that the concept of being human itself has a history.

In sharp contrast, people working in the social sciences and humanities commonly enough think it naive to refer at all to human nature, since it is diverse human identities and their representation – in gender, ethnic groups, sexual preferences, appearances – which appear the decisive things to study. There is a suspicion that reference to human nature hides difference and makes social processes, like possessive individualism, appear natural when they are not. Between this view and the biological and commonsensical view there are large differences and, it often seems, mutual incomprehension. Perhaps, though, the polarisation reflects indifference rather than incomprehension, since so many scholars, of whatever persuasion, take for granted the way of thought, the categories and the terms prevalent in their own neck of the woods; they have more than enough to do to get on in their own field and no time to puzzle about the fact that other fields see things otherwise. Here, however, I do puzzle about why different fields see things differently.

It is necessary to keep in mind that biological theories of human nature and cultural theories of self-shaping are not the only ones in town. There are also religious and humanist believers in 'human' as a spiritual or moral category. Isaiah Berlin once wrote: 'The fact that men are men and women are women and not dogs or cats or tables or chairs is an objective fact; and part of this objective fact is that there are certain values, and only those values, which men, while remaining men, can pursue.'[13] This makes a claim that it is *moral* facts which distinguish human nature. From this point of view, a human science worthy of its name must first and foremost concern itself with action as moral action.

The fourth theme is the explanation of the nature and purposes of 'the history of the human sciences', a phrase which came into use in the last decades of the twentieth century. The phrase is apt to perplex because the human sciences have

13 Berlin, *The First and the Last*, p. 51.

no settled presence in the English-speaking world and because it is not clear what relationship history has to these sciences.[14] But we can explain why there can be no simple definition of the human sciences: it is a field of enquiry in which the very way we shape enquiry is at issue. History belongs in this domain because it is inescapable in human self-understanding.

One further topic, though it is not a theme in this book, requires brief comment. This is the relation between the human sciences and religious thought. Almost every reference to the nature of humankind in the media, as well as much popular writing by scientists, takes this to be the really gripping question. And it is of course the case that a number of religious traditions stress the essential difference of human beings from other entities and hence the essential distinctiveness of knowledge about what it is to be human. Many Christians, including members of both the Orthodox and Catholic churches, maintain that this difference and distinctiveness comes from a God-given soul. When scientists like Dawkins or Pinker argued for an exclusively evolutionary, naturalistic approach to human nature, this belief was their target. Neuroscientists like Francis Crick presented their science as a source of empirical disproofs of the soul and hence of religious belief. Two gross simplifications and misrepresentations often appear in such arguments. Firstly, it is most unhelpful, as well as inaccurate, to subsume discussion about different forms of understanding being human under a supposed conflict of science and religion. What people all too blithely call 'science' and 'religion' are highly diversified and contested realms of human activity; we cannot, without drastic and unacceptable simplification, talk about any kind of one-to-one relationship between two such realms. There are, needless to say, real enough local struggles between groups of scientific and religious believers; the argument over the biology curriculum in schools in the United States is a leading case in point. We may be sure, however, that people who stress conflict in general have an axe to grind (though this of itself does not make them right or wrong). Secondly, more pertinent to my arguments, religious

14 See Smith, *History of the Human Sciences*.

and scientific thought may differ because they are different forms of knowledge rather than because of disagreement about the existence of the soul as an actual entity. Throughout this book I put the argument for the rational existence of different forms of knowledge. The argument, it should be clear, is based in the philosophy of knowledge. There is no argument here for or against religion of any kind. The question of religious belief is for another book, not for this one.

This leads to several other caveats. The book is philosophical, in the sense that its argument concerns historical knowledge and the importance claimed for it. It is, however, philosophical enquiry of a kind that any person can and must engage with if he or she wishes to be reflective about science and human activity. I do not attempt to resolve philosophical questions as a professional philosopher might seek to do. There is, for example, an extensive literature in the philosophy of science, especially influenced by the quantum physics created in the 1920s, about realism – whether scientific knowledge describes the world as it is ('in itself'). Many scientists, like ordinary people, are informal realists; yet there are severe difficulties about holding that well-established areas of science, like wave-particle duality, are compatible with a realist epistemology. It is not possible to go into such matters, even though the arguments made here about different approaches to being human arrive at what appears to be a non-realist position. The problem of scientific realism 'is a profoundly complex metaphysical and epistemological issue', and it is not easy to contribute to it in a new way.[15] There are other topics to get on with, and I do so within the broad philosophical framework which Ian Hacking has described as historical ontology, which studies how objects in the human sciences come into existence with the historical invention of ways to talk about them. 'Categories of people come into existence at the same time as kinds of people come into being to fit those categories, and there is a two-way interaction between these processes.'[16]

15  Alfred I. Tauber, 'Introduction', to Tauber (ed.), *Science and the Quest for Reality*, p. 7.
16  Hacking, *Historical Ontology*, p. 48.

The first chapter looks in more detail at two kinds of systematic knowledge, or sciences, of the human subject, contrasting biology and philosophical anthropology. It also explains the term 'the human sciences'. This is not and could not be a neutral term, and by using it I imply, from the outset, that there are other forms of scientific human self-understanding than biology provides. Many people take this for granted; but many also do not. Though the tradition of philosophical anthropology suggests an alternative, some philosophers, especially Foucault, have decisively questioned its pretensions to inclusive understanding. Foucault dramatised the reasons why it is so questionable to search for definitions or ultimate characterisations of human nature or identity. The reasons are complex and appear to lead to conclusions at odds with the common opinion of natural scientists and lay people alike. The following chapter, on reflexive knowledge, therefore fills out the reasoning. There is, it is argued, no absolutely objective place where we can stand outside what is already one way of life and one way of thought about what it is to be human. What we say about human identity reflects where we stand. By contrast, both biologists and philosophical anthropologists have claimed to provide knowledge from a place outside of history.

It is then possible to discuss more directly the relations between the natural sciences and the human sciences. Is knowledge of human beings strictly comparable with knowledge of the natural world? This question is an old chestnut, but there is scope and need for a re-examination. The reasons which philosophers of the social sciences gave, in the 1960s, for holding that we explain human actions in terms of intentions as well as physical causes retain their force, though recent materialists, especially inspired by neuroscience, gainsay them. It is not, in the last analysis, my purpose to push apart knowledge of people and knowledge of nature; but bringing our understanding of people into continuity with our understanding of nature does not require us to subsume knowledge of the former under knowledge of the latter. Indeed, knowledge of nature needs reinterpretation in the light of knowledge of people, and not vice versa. Much of the detail is historical rather than philosophical. It sets out to show that a conception of a human science independent of biology, but yet still a conception of a *science*, has a long and distinguished history. This history is to be found in the continental European writings of philosophers like Vico, Wilhelm Dilthey and Hans-Georg Gadamer. In the rush for originality, modern readers may feel that earlier terms of debate about the human sciences are passé. That earlier, as well as present, positions are full of flaws no one will doubt. But this earlier literature does create precedent for a rational, scientific search for understanding what is human in terms that go far beyond biology. This is a rich legacy with much more relevance to modern disputes than most participants seem inclined to admit.

The book then turns to examine historical knowledge. I attempt to make good the claim that historical knowledge of belief about what is human is knowledge of being human. The argument requires an account of historical knowledge as narrative, and of narrative as the means for ordering and bringing meaning to the world. Narratives are always for a purpose, and which story we tell about an event or action constructs the world, as it exists for us, in one way rather than another. Writing about being human therefore constructs what it is to be human. The last chapter develops this position, relating it back to the argument about the reflexive nature of knowledge and outwards to the human world of values – the judgements people make about what is good, or beautiful or of significance. I elaborate a version of the argument that there are different forms of knowledge for different purposes. There is a place for historical knowledge as knowledge in the human sciences, and there is an autonomous place for the human sciences in the general scheme of all the sciences. The discussion leads to a suggestion about what in historical knowledge makes it of moral value: it is essential to our notion that others and ourselves have significance or worth.

The epilogue briefly puts the case for historical knowledge of what is thought to be human in more positive and idealistic terms. When Vico proposed a 'new science' in the early eighteenth century, he had in mind a science of human beings making themselves over historical time. At first glance, *self-creation* seems an impossibility – like pulling oneself up by one's own boot-straps. All the same, this book lays out the reasons to think that this is indeed an appropriate way of describing what people do. In consequence, the answer to the question, 'What is human?', lies in the history of the forthcoming answers.

# 1

# Being human

## The human sciences

Knowledge of being human must include knowledge of knowledge-seeking. Before jumping to claims about what human nature is, there is good reason to examine the way in which we organise and state collective self-reflection. It makes a difference, for a start, to talk about 'human nature' rather than about 'being human'. In contemporary English, the former makes an association with the sciences of nature and to the field of what T. H. Huxley, in a book with this title published in 1863, called 'man's place in nature'. The latter alludes to the human condition, the existential concern with the meaning or value of 'being'. It is not that one description is right and the other wrong; but the choice has consequences.

There is an apparently straightforward naturalistic way of stating that objective knowledge of being human requires a science of human nature, parallel to the sciences of physical nature. Biologists like E. O. Wilson have confidently asserted that knowledge of evolution has established the basis for this science. But this commonplace reference to 'human nature' is not neutral. In order to show this, it is useful to return to the words 'nature' and 'science'.

'Nature' is famously polyvalent. When Hume discussed the basis of morals, he first noted 'that our answer to this question depends upon the definition of the word, Nature, than which there is none more ambiguous and equivocal'.[1] There are three principal denotations, which common reference to human

1 Hume, *Treatise of Human Nature*, p. 474.

nature runs together. Firstly, 'nature' denotes everything taken to be part of the sensible world. This usage acquired its original force in the contrast which religious people drew between the natural or sensible world and what they believed to be not part of that world – the supersensible world of the Divine or of spirits. This usage, therefore, is associated with ancient debate about whether human reason and human aspiration for 'the good' exists in nature or in opposition to nature. For the many people in the modern world who believe that there is nothing beyond nature with which to contrast nature, humankind is by definition part of nature, and so is everything that it does and produces, like cities or technology or the arts. It is all natural. Thus Whitehead observed: 'It is a false dichotomy to think of Nature *and* Man. Mankind is that factor *in* Nature which exhibits in its most intense form the plasticity of nature.'[2]

Secondly, 'nature' denotes the essence of something, that which makes something what it is. This is the usage in discussions about such things as God's nature, the nature of goodness and the nature of human nature. It has an old-fashioned ring as it conjures up associations with Platonic and Aristotelian attempts to understand the world in terms of pure forms or essential natures. Christian writers tirelessly discussed the nature of the soul in this sense. There has been a more or less continuous debate about whether these forms, essences or natures are real or whether our descriptions denote products of the intellect. In everyday speech, ordinary people and scientists alike are most often realists about this, and they assert that it is in the nature of something to act as it does. Cat owners dull their sensibility with the observation that it is a cat's nature to kill birds.

Thirdly, 'nature' denotes a contrast between what people feel is real and society, culture or artifice. This usage has a long history in the contrast between what is natural to people and what is an art, and in the contrast, so important to seventeenth- and eighteenth-century political thought, between the state of nature and the state of civilisation. The human sciences in the

2 Whitehead, *Adventures of Ideas*, p. 99.

twentieth century were preoccupied with relations between nature and nurture (to use Francis Galton's terms). The word 'culture', though notoriously hard to define, became ubiquitous in order to effect a contrast with 'nature' understood in this sense.

Awareness of these different meanings makes it possible to see just how diverse notions of human nature are. The term may refer to a metaphysical, existential or religious belief about being, the life-world, existence or the human *telos*. Or it may refer to a physical inheritance from the life of human-like animals in the Pleistocene age. Or it may refer to shared historically formed, and history-forming, activity. A reference to human nature by itself says nothing precise. And when people claim exclusive knowledge of human nature, they play on the multiple meanings of 'nature', implying, for example, that a particular view of what is natural (not social) to human beings is their (essential) nature.

All the same, in current English the term 'human nature' is particularly associated with biology: the biological way of thought about human nature appears natural (to play on the word). This is because, in secular western thought, the word 'nature' is so closely tied to a particular form of knowledge which equates what exists with what the natural sciences describe and explain. Those who claim that natural science knowledge has unique standing as knowledge treat the nature of something as the same as the nature which natural scientists ascribe to it. Natural scientists, however, are not engaged in simply observing how the world is, since, in Neil Evernden's words, 'There *is* a metaphysic lying behind the simple existence of the word *nature*. It is not simply a description of a found object.'[3] The same is true for 'human nature'.

All human activity, from language to law and from pop to politics, is a product of human nature. It is this activity which we usually – if loosely – denote as culture. Grasping for simplicity, debate sometimes proceeds as if there is a stark choice between two positions, the first maintaining that human nature is nature – and hence a subject for the natural sciences, and

3 Evernden, *Social Creation of Nature*, p. 21.

the second maintaining that human nature is culture – and hence a subject for the human sciences. This cannot be a viable way to carry on, since, of course, in a certain sense culture is as much a part of nature as anything else. We must grapple with complexity.

I now turn to the notion of science and to the well-known but still inadequately appreciated difference between English-language and continental European usage. 'Science' and 'natural science' are commonly synonymous for academic and non-academic English-speakers. This certainly signifies the standing and authority the natural sciences have as the model for knowledge. As a result, however, English usage makes it appear as if the history of science, including the history of the human sciences, is by definition a history of the assimilation of knowledge to the natural sciences. Usage also makes comprehensible the judgement that certain disciplines – history or literary theory, for example – since they are not natural sciences, are not sciences at all. It explains why there have been recurrent Anglo-American debates about whether the psychological and social sciences really are sciences, predicated on the assumption that if these activities are strictly comparable with the natural sciences they are sciences; if not, not.

This English-language usage is relatively new. It seems to have come about as a result of changes in the social organisation of knowledge in the nineteenth century, changes which created science, understood in the way people subsequently have used the word, as a set of specialised disciplines. Earlier writers used the word to refer to *any* systematic body of learning held to be true. William Whewell, who in 1833 introduced the word 'scientist' to describe a specialist in some branch of knowledge, was himself knowledgeable across a vast field and continued, in rather an eighteenth-century way, to hold that in principle the study of the moral and social sphere, as well as the natural, belongs to science.[4] According to the *Oxford English Dictionary*, it is only in the second half of the nineteenth century that the word 'science' began to contract and to denote natural science.

4 Yeo, *Defining Science*, pp. 24–5.

The older usage continued in continental Europe. In Germany, Russia or Sweden, the university disciplines such as literature, history, geography, linguistics, art history, and even – in some settings – theology, as well as physics, chemistry and biology, are sciences. In medieval and early modern universities, '*scientia*' denoted knowledge grounded on principles or presuppositions known to be true, and hence science was any true body of knowledge. ' "*Science*," as a philosophical term, is the clear and certain understanding of something through principles either self-evident or demonstrable.'[5] The word 'science' differentiated knowledge from opinion. Theology was certainly a science, even the most basic science. To decide whether a particular discipline – history, for example – is a science involved a judgement whether its knowledge is rationally grounded and true, not whether it is a natural science.

In practice, the picture is more complex, since calling a field a science asserts that it does possess true and rational knowledge. The word 'science' is normative as well as descriptive. The use of the word is therefore associated with arguments about whether areas of research do have the cognitive authority which they claim. Thus, in France, purists have sometimes held that, as a matter of fact, only the mathematical physical sciences have the rational and formal structure which qualifies a field to be a science. Even discounting that rather extreme position, it is still the case that *l'histoire des sciences* generally denotes the history of the natural sciences, not including *les sciences humaines*. This custom signals that, though there is systematic knowledge in the human sciences, it is not thought to meet the same epistemological standards as the natural sciences and should not therefore properly be called scientific. When the philosopher of science Georges Canguilhem, in an influential lecture in 1956, questioned the scientific standing of psychology, his point was twofold: psychology, though stated to be a unified scientific discipline, has not been able to specify the unity of the object which it purportedly studies; and psychology does not have grounds for the authority which

---

5 Diderot's *Encyclopédie*, quoted in Kelley, 'Problem of knowledge and the concept of discipline', p. 18.

it exercises as a human technology. His polemic was that psychology has not met certain epistemological and moral standards, not that psychology has failed to become a natural science.[6] Over the next decade, many French intellectuals accepted Lacan's claim that psychoanalysis is the basic human science, though Lacanian theory was remote indeed from what English-speaking natural scientists imagined a science to be.

It is therefore open to debate what the scope of the history of science should be. As historian of *les sciences de l'homme*, Georges Gusdorf, once wrote: 'L'idée de science est une variable historique.'[7] There is little intellectual mileage in trying to draw a sharp boundary, in general, between what is science and what is not. In the late medieval period and Renaissance, for instance, scholars listed among the sciences such subject areas as music, arithmetic, optics and astronomy, but also metaphysics and theology. Pascal's anxious query was about what sort of science theology could be, not whether it was a science: 'Theology is a science, but at the same time how many sciences?'[8] As late as 1829, the historian Thomas Babington Macaulay referred to 'that noble Science of Politics'.[9] After all this, Bruno Latour's pronouncement, ' "science" – in quotation marks – does not exist', is a temptation.[10]

It is now possible to turn to the notion of the human sciences. Just as with words to describe being human, words to describe the field of study of what is human prejudge the subject. Eighteenth-century writers, like Hume, referred to 'the science of man', while Hume's French contemporaries used the expression *'la science de l'homme'*. But in modern English, 'man', even when denoting people in general, has connotations of one gender rather than the other, and it is therefore widely felt that using 'man' as a generic term makes the implicit, and

6 Canguilhem, 'Qu'est-ce que la psychologie?'; see Braunstein, 'La critique Canguilhemienne'.
7 Gusdorf, *De l'histoire des sciences*, p. 15.
8 Pascal, *Pensées*, p. 18 (§ 65).
9 Collini, Winch and Burrow, *That Noble Science of Politics*, p. 100.
10 Latour, *Pasteurization of France*, p. 216.

unacceptable, claim that male characteristics are normative for everyone. Interestingly, the word 'man' has been thought to be gendered in a way that *'homme'* in French, or *'Mensch'* in German or *'chelovek'* in Russian have not. Quite why this is so is a puzzle, but it is so, and as it is clearly the business of the human sciences to study the place of gender in knowledge, it is unacceptable to refer to such study as 'the science of man'.

The term 'the human sciences' came into use as the best available alternative. The social fact, however, is that the term does not have a settled or agreed meaning. This is not a matter of sloppiness or confusion but reflects genuine differences of view about what sort of scientific knowledge of humans is possible. Bruce Mazlish suggested distinguishing 'the singular, *human science*, [which] is specifically the study of human evolution as it moves increasingly from the physical to the cultural . . . [from] the plural, *human sciences*, [which] although it also includes the first, embraces all disciplined efforts to understand specialized aspects of human behavior – political, economic, sociological, and so on – not excluding the humanistic approaches'.[11] Though not everyone will agree to this distinction, it is indeed different views about the relation between approaches in the humanities, like history and literary theory, and biological approaches, which makes for different usage. I will use the term, in the plural, in Mazlish's sense, but, by doing so, also signify that it is a field which examines its own ways of defining its subject. The human sciences shape themselves through debate about what the subject is. Thus, for example, the human sciences are not the same as 'the behavioral sciences' – the American spelling appropriately indicates their provenance – because the former question 'behavior', as the latter do not, as a defining category. Some authors, indeed, 'employ the term "human science" . . . to embrace the full range of disciplines required to make sense of the human condition'.[12]

11 Mazlish, *The Uncertain Sciences*, p. 8.
12 Roberts and Good, 'Introduction' to *Recovery of Rhetoric*, p. 2.

It does not need argument to establish the fact – whatever some scientists think should be the case – that there is at present no unified human science, in the singular. In the eighteenth century, *'la science de l'homme'* and 'the science of man' were used in the singular form. But even in the Francophone world, where the earlier expression is still a living one, it is now used in the plural form – *'les sciences de l'homme'*. There was once hope for the unity of knowledge, less resignation to knowledge being parcelled out between a host of disciplines. A conventional list of these disciplines would include the social sciences, economics, cultural studies, the psychological sciences, geography, human biology, anthropology, political science and linguistics. This, however, is a minimal list, easily expanded by including, for example, literary studies, religious studies and the management sciences. I shall claim that history is a discipline in the human sciences. Which disciplines, as a matter of fact, get included in the human sciences is a question of local, often administrative, convenience, or of local power-play among academics.

The goal, however, is not to define a field by listing disciplines. The intellectual importance of the generic name is to indicate that claims to knowledge about the human subject pose a common core of problems, and that it is therefore productive to keep open communication across the full range of potentially relevant pursuits. I would also claim that there are profound philosophical and moral reasons why knowledge about the human subject, even the most basic characterisations of what the subject is, should remain open. The human subject has not been given once and for all, and we do not know its bounds.

All sorts of specialists claim knowledge of people. This is the obvious reason why the collective expression 'the human sciences', in the plural, is useful. The term, in itself, neither includes nor excludes any one particular body of knowledge. The term encourages thought which does not prejudge, as research conducted within the framework of a single academic discipline necessarily does, what sort of knowledge will command authority. As the term has no clearly defined and no universally accepted referent, it leaves it open for institutional and intellectual interests to shape the field for local purposes.

This is appropriate in an area as complex as the human sciences. 'We need to set ourselves free to make such connections and disconnections between fields of enquiry as seem appropriate and productive, not to prejudge what may be learned from what, what may traffic with what.'[13]

What the subject *is* in studies of being human has itself been part of what is at issue. This continues to be so in the modern human sciences. Activity in the human sciences crosses the conventional Anglo-American division between the humanities, the social sciences and the natural sciences. Even in France, where *les sciences de l'homme* are institutionalised, there is still argument, as the history of psychology shows, about what they include and where to place them. In the German-speaking world, discussion has frequently turned on the meaning and relations of *Geisteswissenschaft* (literally, the 'science of spirit', though the term is well known for translating awkwardly) and *Naturwissenschaft* (the 'science of nature'). All the same, though usage is conventional, language and decisions about a discipline's institutional home both reflect and confirm commitment to one rather than another theory of knowledge.

It is only since the 1960s that the term 'the human sciences' has become at all common in the Anglophone world. One reason for its spread is the influence of translations from the French, translations which seemed to offer new ways of thinking about the human subject. Most prominent was the translation, in 1970, of Foucault's *Les Mots et les choses* as *The Order of Things: An Archaeology of the Human Sciences*. The main reason for the term's use, however, is probably simply the awesome amount of investigation into human life in many, many disciplinary areas. Each area is in itself so large that those who work one patch hardly feel the need, and do not have the time, to look over the fence at what is going on next door. If they do, often they will not understand. Reference to the human sciences has a manifest practical purpose in this bewildering situation and appears simply a convenient informal term. For example, Perry Anderson's well-known

13 Geertz, 'The strange estrangement', p. 89.

survey of British culture referred to the human sciences, in which he included philosophy, political theory, history, economics, psychology, psychoanalysis and aesthetics, without comment on the term.[14] Nevertheless, other people have not used the term, and the term 'social sciences' remains, in a number of contexts, more recognisable as a collective term for research.[15]

Convenience also seems to explain use of the phrase 'the history of the human sciences'. The disciplines in the social sciences, at least in their institutionalised form, are modern, and it is helpful to have a term for research on their collective history, especially in the period before there were modern disciplines. It is greatly preferable to work under the rubric of a collective term than to engage in the practice, still common in histories written by scientists within particular disciplines, of projecting back the modern discipline into the past, producing what the historian of English culture Stefan Collini called 'tunnel-vision'.[16] Writing under the heading of the history of the human sciences supports the open-ended study of the period before the modern disciplines existed, in a way that does not take discipline boundaries as given. The term is, without question, anachronistic. All the same, it helps make space for historical research sensitive to earlier classifications of knowledge and to the roots of the classification now in use. The historian Anthony Pagden, for example, wrote, referring to the early modern period: 'Since the formal division of the social sciences into their modern faculties takes place outside the historical period with which I am concerned I shall frequently use the more comfortable term "human sciences" to describe those areas of inquiry whose subject is human social behaviour.'[17] Rather than tracing early modern anthropology as if it were then a discipline, he searched, in the areas of learning he thought relevant, for the constitution

14 Anderson, 'Components of the national culture'.
15 Porter and Ross, 'Introduction: writing the history of social science', pp. 1–3.
16 Collini, '"Discipline history" and "intellectual history"', p. 391.
17 Pagden, 'Eighteenth-century anthropology', p. 224.

of the objects of what subsequently became anthropological knowledge.

I want to argue at length, however, that there is more to using the term 'the human sciences' than making oneself 'comfortable'. As I have already stressed, no description of human subjects is neutral. For one thing, as the philosopher Joseph Margolis noted, 'to address the *human* sciences is to confront ourselves with the possible discontinuities between the "human" and the "natural" sciences'.[18] For another thing, classification of knowledge has changed, and may change again; and to describe work as the history of the human sciences signals that enquiry will extend into classification and concepts. Nothing is given. In Charles Taylor's words: 'A study of the science of man is inseparable from an examination of the options between which men must choose.'[19]

### Biology and human identity

Asking the question 'What is human?' invites an answer in terms of essential attributes – a claim about human 'nature'. The Judeo-Christian world, over many centuries, maintained that a transcendent principle, a creative act of God, is the source of this real nature and identity. This is indeed the background to the modern notion of a person, a notion taken, for Christians, from the manner in which Christ lived in human form. The Orthodox and Catholic churches hold to the faith that each individual human being has an essential nature, a God-given soul. Within this framework of thought, if there is to be a human science, it must, at base, be a theological science with knowledge of the soul. Hence for Catholics the immense importance of Aquinas's systematic examination of the nature of the soul. Following the break-up of the Soviet Union in 1991, there have been Russian attempts to re-create scientific psychology, in conformity with Orthodox tradition, as the science of the soul. There have also been non-Christian visions of psychology as the science of the soul, understood as a spiritual

18 Margolis, *Science without Unity*, p. xv.
19 Taylor, 'Interpretation and the sciences of man', p. 54.

principle inherent in human nature. C. G. Jung, to take a notable example, re-expressed belief that man has a soul, and he did so in terms which inspired many people who found the articles of Christian faith incredible. For Jung, humans do have an essential nature, which he characterised in terms of the archetypes of the collective unconscious, the content of his human science.

By contrast, many natural scientists think that evolutionary theory is the core of the human sciences. This is a controversial area to venture into, and it is certainly not my purpose to survey evolutionary theories of human nature and their strengths and weaknesses. What I do need, however, are arguments to show that the evolutionary approach cannot be either the only understanding of what it is to be human or the underlying basis of all other ways of understanding. It must be accepted that opinion is at present markedly divided over such questions. Some evolutionists will think I simply have not grasped what science has shown to be the case. Other people, in complete contrast, will think that there is no need to erect a defence of the obvious.

Modern evolutionary accounts of human origins continue to reflect belief that there is an essential human nature, the nature all people share through their common root. In this respect, at least, modern science and Judeo-Christian belief share much in common. In the evolutionary story, human nature originated in nature not in a special act of divine creation, but the story, like the Christian story, pinpoints a common inheritance. Mary Midgley may well have been right: evolutionary science has acquired a social function once carried out by religion, providing persuasive answers to questions about human identity in the form of a story of a shared origin.[20] The conditions of impermanence in modern life perhaps prompt people, faced by a sense of loss, the absence of purpose and the presence of risk, to seek some permanent ground beyond the ephemeral. This may be one reason for the huge appeal of evolutionary accounts of human nature at the beginning of the twenty-first century. Here, in a biological nature

20 Midgley, *Science as Salvation.*

fixed several million years ago, surely, lies a common humanity, a touchstone for human affairs. Yet it is questionable how
much such thought has to offer to those who search for direction in life or feel lost amidst the re-creation of identities
around them. While certain evolutionary notions of purpose
and identity satisfy particular individuals or groups, at least
for a time, they do not have authority more widely, as the
flourishing of religious alternatives testifies.

There is a certain irony in evolutionary theory becoming the
source of claims about the human essence or human nature.
The evolutionary biologist Ernst Mayr, for one, stressed that
it was Darwin's great intellectual achievement to have broken
with the essentialism, or belief in each thing having an essence
and each species having a fixed nature, which was so central
to European thought passed down from the ancient world. 'It
was the [evolutionary] study of diversity more than anything
else which undermined essentialism, the most insidious of all
philosophies. By emphasizing that each individual is uniquely
different from every other one, the students of diversity focused
attention on the role of the individual.'[21] The central insight of
Darwin's vision of nature was the *diversity* of species and of
individuals within a species, not the existence of an essential
core of particular characteristics. Darwin's work made it an
empirical matter to determine which characteristics, varying
over what range, individual humans actually share, not to
seek to describe certain characteristics as essential and to
use those characteristics to define a norm for what it is to be
human. Many evolutionary accounts, however, claim that long
ago ('once upon a time') in the Pleistocene age a particular set
of traits became fixed, that they have not subsequently changed
and therefore that there is an essential human nature. Some
writers have then gone further and conflated what they think
essential with a norm or a standard. John Tooby and Leda
Cosmides, in their introduction to evolutionary psychology,
used the example of jealousy: 'the most illuminating question
is whether every human male comes endowed with developmental programs that are designed to assemble . . . evolutionarily

21 Mayr, *Growth of Biological Thought*, p. 249.

designed sexual jealousy mechanisms that are then present to be activated by appropriate cues'.[22] This presumes a social norm – in this case, male jealousy – and then explains it as a trait fixed by the life of early humans on the savannah. The writing accords jealousy status as human nature, and it makes this nature available as a standard against which to judge the actual life of people, however diverse and flexible that life may actually be.

It is helpful here to list four critical approaches to the claim that evolutionary science is authoritative in knowledge of 'the human'. Later discussion enlarges on these themes. There is, first, the epistemological problem. This is the difficult matter of clarifying how a scientific theory constructed by reflective consciousness can be held to be an explanation of reflective consciousness. The view taken here is that any scientific theory presupposes certain epistemological commitments which that theory itself cannot validate. There is always scope for another level of understanding than the one actually embodied in a particular view of the world. The second layer of criticism restates the argument, familiar from the philosophy of the social sciences, that human actions require explanation by reference to inherently social rules and intentions. No amount of biological information can provide this kind of explanation. The third layer concerns values. It just is the case that people differentiate the world in terms of significance, beauty, goodness and truth. It is one thing to claim that these judgements have a biological function and serve adaptation and survival; it is another thing to say that the biological function constitutes their meaning. Is what we mean by truth, for example, a state of correspondence between an organism's knowledge and its environment which helps an organism to survive? Should we define 'the good' as that which, as a matter of fact, enables an organism to reproduce more of its own kind? Rather, evaluative statements are human, historically local acts. There is no more value in survival in itself than non-survival. The fourth layer of analysis examines the quality of the empirical claims evolutionary biologists make about

22 Tooby and Cosmides, 'Psychological foundations of culture', p. 45.

how human evolution has actually occurred and what legacy
it has left in the way people experience and act in the world.
All these layers of criticism do not deny the statement with
which Mazlish began his study of the human sciences, that
'our context for such knowing must be evolutionary', but they
do put that 'context' itself in context.[23]

As a way into these questions, and as an illustration of the
complexity which popularisers of the biological approach are
apt to gloss over, I return to the question of essentialism in
Darwin's own writing. When Darwin wrote about morality – a
commitment to which, as a Victorian gentleman, he took for
granted – he did imply that he believed in one standard. This
is one kind of essentialism. Moreover, he took the realisation
of this standard to be human progress. It was, for him, a
matter of fact that there has been moral progress, though in
retrospect we may think this says more about his particular
Christian culture than a universal achievement. 'He insisted
upon the primate origin of human beings, but allowed them
thereafter to enjoy a good Victorian Whig destiny, fighting
their way up a ladder of moral improvement using the weapons
of hierarchy, order and education.'[24]

When Darwin worked up his argument for human evolution, after he had published on the origin of animal and plant
species, he wrote with critics in mind who believed that human
beings alone possess certain mental attributes. Darwin very
effectively opposed this kind of essentialism. In *The Descent of
Man* (1871), he listed the capacities of reason, moral sense,
religious feeling, aesthetic appreciation and language, all of
which opponents of belief in human evolution had claimed
unique to human beings, in order to show the variety and lack
of agreement among his critics. 'I formerly made a collection of above a score of such aphorisms [putting up a barrier
between humans and animals], but they are not worth giving,
as their wide difference and number prove the difficulty, if
not the impossibility, of the attempt.'[25] This was skilful rhetoric

23 Mazlish, *The Uncertain Sciences*, p. 1.
24 Kuper, 'On human nature', p. 289. Also Ruse, *Monad to Man*,
   pp. 150–69.
25 Darwin, *Descent of Man*, p. 49.

as it prepared the ground for his own argument that each attribute could, in however elementary a form, be found in animals. He persuasively pointed out that if we can see continuity in mental life between animals and humans, then we may believe that humans, like animals, have evolved by natural selection. Entirely reasonably, given his evolutionary argument, he read human traits into animals and animal traits into humans.

Darwin did not write about the soul or the immortal principle, and he carefully tried to step around religious sensibilities, not least those of his wife. For himself, he observed only: 'I fully subscribe to the judgment of those writers who maintain that of all the differences between man and the lower animals, the moral sense or conscience is by far the most important . . . it is summed up in that short but imperious word *ought*, so full of high significance.'[26] Nonetheless, he went on to outline a complex hypothesis about how elements of a moral sense began to evolve in the social life of animals and human predecessors. Darwin, therefore, stressed the continuity of animal and human worlds and at the same time stressed the unique moral progress which his Victorian, human world had actually made. The result was a picture of an essentially moral humankind even though the roots of morality lay, in his view, in the chance diversity of early human traits.

It is certainly correct that the religiously committed British writers on human nature, whom Darwin had in mind as his opponents, picked on particular attributes, like the moral sense and the power of reasoning with abstract universal concepts, as empirical proof of human uniqueness. Such writers argued from this uniqueness to a special origin. Thus when Darwin wrote in relation to the moral sense, that 'this great question has been discussed by many writers of consummate ability; and my sole excuse for touching on it is . . . because, as far as I know, no one has approached it exclusively from the side of natural history', he was disingenuous.[27] As he well knew, it was the existence of ways of knowing the human world other

26 *Ibid.*, p. 70.
27 *Ibid.*, p. 71.

than from 'the side of natural history' which was the substantial matter at issue. It still is. In practice, indeed, as I have just argued, Darwin's belief about what is good did not derive from natural history; he took for granted values derived from his society. He then attempted to say that the basis for these values had been laid down in the evolutionary process. This involved him, as it involved later evolutionists, in identifying what is good, ultimately, with what as a matter of fact has survived as a result of natural selection. But this contradicted his Victorian notion of what is good, which clearly derived from a historically formed Christian culture.

Darwin's intellectual and moral position, and its contradictions, continues to be widely shared. Many people want to trace the roots of moral action to evolution but nevertheless to argue that only certain kinds of action, those kinds which most people as it happens accept are good, are good. The way out of this contradictory position, in my view, is to recognise that the act of judging what is good can take place only in and apply to a historical culture. There is no meaning in saying there is good or bad 'in nature'.

This is a view which does differentiate humans from animals, since only humans have a historical culture and, hence, also moral culture, whatever humans may, as a matter of fact, also share with animals. The central intellectual challenge, then, it would appear, is to understand the evolution of reflective consciousness and language, and hence the evolution of culture. But it is actually possible to turn this challenge around, in a way which biologists find hard to sympathise with or, perhaps, even see as a rational thing to do. It is possible to ask, not how did evolution give rise to conscious reflection, but how does the existence of conscious reflection make it possible to have knowledge (including knowledge of evolution). It is very important that the former question appears to have the possibility of an empirical answer (only the wildest supporter of evolutionary biology would assert that we already possess the empirical answer). The latter question appears to leave the matter with philosophy, a field where few people expect definitive answers. But actually this is a false contrast. The evolutionists, too, take a philosophical stance, presuppose a theory of knowledge and adopt a view about the relation of reflexive consciousness to its subject matter. It is just that the empirical way of thought has become so taken for granted it does not look this way.

The literature on neuroscience has included a number of influential attempts to make the whole problem of reflective consciousness a scientific one and, as such, a problem which, sooner or later, empirical knowledge will solve. From this point of view, 'the ultimate question is: How exactly does nervous tissue cause consciousness?'[28] And, in Susan Greenfield's upbeat assessment, 'hard experimental data may be just around the corner'.[29] Progress in knowledge of the brain, supported by computer modelling and cognitive psychology, has greatly encouraged belief that science is about to reveal, in material terms, what consciousness really is. A number of influential philosophers of mind have offered support. These philosophers, like Paul M. Churchland and Owen Flanagan, critical of the earlier separation between philosophy and neuroscience as different kinds of activity, have argued that 'mental processes just are brain processes' seen in a certain kind of way.[30] In the future, they argue, the scientific way of thought will replace any reference to mind by statements about brain, and the question about reflective consciousness will be a question about brain states. It must be said, however, that no one at present has any idea how a brain could have such a function. Even a confident proponent of evolutionary psychology like Pinker, speaking of the brain's production of sentience, or the qualitative feel of consciousness, has said: 'I have...no idea of how to begin to look for a defensible answer. And neither does anyone else.'[31]

These materialist approaches to consciousness, like the evolutionary approaches to knowledge of what is human, in as far as they claim to have exclusive or basic authority, fall foul of the difficulties I have already been raising: the epistemological problem, the social nature of explanations of human

28 Greenfield, *Journey to the Centers of the Mind*, p. 12.
29 Greenfield, 'Soul, brain and mind', p. 124.
30 Flanagan, *Consciousness Reconsidered*, p. xi; also Churchland, 'Eliminative materialism'.
31 Pinker, *How the Mind Works*, p. 146.

action, the question of values and the quality of the empirical evidence. Philosophers like Dennett, who wrote *Consciousness Explained* (1991), think they have answers. But the rest of this book will build up the case as to why we cannot translate, and shrink, the problem of human self-understanding into an empirical, natural science theory of the brain's production of consciousness (even if we had such a theory).

Darwin himself left the topic well alone. When he wrote about consciousness he simply gave a great quantity of evidence that animals too exhibit it, and here he had the weight of common opinion on his side. This was important for what he most wished to stress, animal–human continuity. He was also aware of – even if he did not get to grips with – the argument that human consciousness has, through language and culture, created a form of life that did not exist earlier in nature. His co-theorist of natural selection, Alfred Russel Wallace, had drawn his attention to this, though Darwin greatly feared that Wallace's way of putting the point offered arguments to the opponents of belief in human evolution. Wallace, in 1864, wrote, referring to the evolution of human rational consciousness: 'A being has arisen who was no longer necessarily subject to change with the changing universe – a being who was in some degree superior to nature, inasmuch as he knew how to control and regulate her action.'[32] This was a major insight, though it was left to later generations of writers on evolutionary thought to follow it up. It is not possible to discuss 'man's place in nature' independently of the cultural activity through which nature itself has become an object of knowledge and manipulation. It suggests that it is actually not of decisive importance, for understanding what possibilities for action people now have, to settle whether animals do or do not have certain mental attributes like reason and language, or to speculate about what all people have inherited from their Pleistocene predecessors. Even if we were to conclude that animals have some element of reflexive consciousness, we still know the difference between humans and animals. Even if life on the savannah has left a legacy, we still know the difference between

32 Wallace, 'Origin of human races', p. 52.

early and modern people. The difference is the reflexive, historical culture, including biological knowledge, that humans have established and animals and early humans do or did not have. If there are biological capacities in common among people, and it is an almost empty truism to accept this, it is not possible to describe these capacities independently of their expression in particular forms of cultural life. The capacities are embedded in the way life is lived; they do not exist independently of that life, except as abstractions in the theories which biologists and psychologists formalise according to their scientific way of life. As the anthropologist Clifford Geertz once observed, 'there is no such thing as a human nature independent of culture'.[33]

There has been extensive intellectual opposition to the exclusively biological view of human identity. Through the middle decades of the twentieth century, for example, a non-biological science, structuralism, conceived as a search for a temporal formal relations of the basic units of human (mental, linguistic or cultural) activity, much influenced linguistics and social anthropology. I shall, however, focus on philosophical anthropology. I use the term rather broadly to encompass belief that there is something given in reason or morality itself that marks out what is essential to being human, something beyond both biology and history. One common form of such belief emphasises that humans, uniquely, have reflective consciousness – they not only know but know that they know. The outward sign of reflective consciousness is the world of symbolic representation, especially language, and the forms of collective life symbolisation makes possible. I will give a historical sketch of philosophical anthropology and then indicate why this attempt to define an essential core of being human has many troubles.

## Philosophical anthropology

At the beginning of his student lectures on logic, Kant laid out the key purposes of philosophy:

33 Geertz, 'Impact of the concept of culture on the concept of man', p. 49.

The field of philosophy . . . can be brought down to the following questions:

1. *What can I know?*
2. *What ought I to do?*
3. *What may I hope?*
4. *What is man?* [*Was ist der Mensch?*]

*Metaphysics* answers the first question, *morals* the second, *religion* the third, and *anthropology* the fourth. Fundamentally, however, we could reckon all of this as anthropology, because the first three questions relate to the last one.[34]

Kant here used the word 'anthropology' as a collective term for enquiry through experience into being human, including enquiry into knowledge, morals and religion. This usage was already common in the German-speaking academic world of the 1760s, in the period before Kant made his critical turn in philosophy, when it indicated an interest in deriving knowledge about humankind from experience rather than logic, pure reason or dogmatic theology.[35] Later, Kant tended to employ the word in a more restricted sense to denote pragmatic anthropology, knowledge concerned with individual human character, its control and improvement. In this connection, he gave an annual course of lectures, which were infinitely more accessible than those on logic, which covered such matters of common interest as perception, sleep and manners.[36]

By opening lectures with the question, 'What is man?', Kant implied the possibility of a unified answer. When, after 1770, he initiated a kind of Copernican revolution in philosophy, the range and depth of his thinking led, however, not to ready formulas but to three major critiques, and it required great subtlety on his part to show how they formed a unified system. In Kant's philosophy, the human soul, though in principle a unity, has in its nature the three powers of cognition, of feeling pleasure and pain and of desire. In his three critiques he

---

34 Kant, *Kant's Logic*, p. 538.
35 See Zammito, *Kant, Herder, and the Birth of Anthropology*, pp. 292–302; Clark, 'Death of metaphysics'.
36 Kant, *Anthropology from a Pragmatic Point of View*.

examined the necessary logical or rational (not contingent and material) conditions that must obtain if the knowledge that we have with these powers is to be possible. These critiques, he claimed, are the most fundamental possible logical analysis of the three higher powers of the soul as they act in cognition – in the cognitive powers of understanding, judgement and reason.[37] The analyses, in turn, make possible comprehension of rational knowledge of nature, of art and of freedom (and morality). Kant's writing in the critiques was dauntingly complex and obscure, making it difficult for any non-philosopher. But his contemporaries and later commentators thought he had brought about a revolution in philosophy because he analysed the form that knowledge necessarily has if, as it does, it gives order to human life and experience, rather than expecting human life and experience to find that order in the world. Human cognition in ordering the world is, for Kant, the way that humans constitute themselves as truly human. By arguing that in cognition we do not have access to things in nature, including the soul, 'in themselves', but to the forms of things that cognition necessarily gives them as a condition of knowing them, he made human action the centre of all knowing. This refounded the project of anthropology, understood in the largest sense. It also bequeathed a lasting preoccupation, to use modern terms, with how human subjectivity relates to the sciences. I shall return to this again and again from different directions.

In his discussion of the power of cognition associated with desire, Kant, nevertheless, did conclude that there is knowledge of a moral imperative belonging to a supersensible reality (a reality known by reason not experience) 'in itself'. Moreover, he held, in respect of this imperative, human beings have an essential, necessary freedom. In writing for a wider audience, he then argued that humankind only gradually acquires this freedom, as a matter of contingent fact, through cultural development. There is a potential in men and women which history turns into reality. This very influential argument

---

37 See the introductory summary in Kant, *Critique of Judgment*, pp. 37–8.

was a reference point for what 'enlightened' thought was then understood to be. There is a 'germ on which nature has lavished most care – man's inclination and duty to *think freely* ... [which] gradually reacts upon the mentality of the people, who thus gradually become increasingly able to *act freely*'.[38] This ideal of freedom was to remain central to later philosophical anthropology.

Anthropology also inherited from Kant the problem of specifying how such human freedom could act in a world of mechanistic events and necessity, the world the cognition of which Kant had analysed in his first critique as part of the analysis of reason in general. Kant dealt with the question in the third critique, the critique of judgement (*Kritik der Urteilskraft*, 1790), which, he intended, would 'throw a bridge from one domain [nature] to the other [freedom]'. He thought this could be done through the analysis of judgement, the form of cognition dependent, in the empirical sense, on the power of feeling pleasure and pain. It led him into an account of the 'purpose' that the cognising mind necessarily determines is part of knowledge of an object experienced as beautiful or part of knowledge of living things. As he stated, referring to the rational (not experiential) 'concept of a purposiveness of nature', it is 'the one and only way in which we must proceed when reflecting on the objects of nature with the aim of having thoroughly coherent experience. Hence it is a subjective principle (maxim) of judgment.'[39] With this established, Kant then proceeded to use it as the 'bridge' or 'transition' to the argument that our knowledge of necessity in nature and of freedom in human action can be reconciled as different forms of knowledge of a supersensible reality.

The success or otherwise of Kant's project is for specialists to determine. What matters here is that he set out the task which remained pivotal for later philosophical anthropologists: relating what we know through experience and natural science – 'man' as biological being – with what we determine

38 Kant, '"What is enlightenment?"', p. 59.
39 Kant, *Critique of Judgment*, pp. 36, 23.

to be the logical conditions of knowing anything at all. This might be described as the question of the relation between embodied and reflexive knowledge.

Kant's analysis led him to posit teleology, the purposive relation of parts to whole, as a necessary principle of our reasoning about life, including human life. Expressed in anthropological terms, being human is activity in the light of goals or ends. German writers in the generation after Kant described this activity as a human self-forming, or '*Bildung*', a word which came to have almost sacred status in idealist writing about the purposes of education in the broadest sense. An emphasis on goal-oriented forming remained a distinguishing feature of writings in philosophical anthropology, which thereby rendered human values intrinsic to the *nature* of being human. This was, it scarcely needs saying, a way of thought increasingly separated from the main mechanistic strands of the twentieth-century life sciences, in particular, molecular biology, natural selection theory and neuroscience. In Kant's age, however, and in fact well into the twentieth century in German biology, principles of teleological reasoning continued to link natural science and anthropology.

Kant's belief that purposes are inherent in the logical scheme of human understanding was especially evident in a sketch for a universal history – a 'philosophy of history', in contemporary terms – which he sent to Berlin's leading intellectual journal. He laid out nine propositions which, he intended, make clear the place enlightened progress has in the historical process. The first, basic proposition stated that there is a purpose in nature: 'All the natural capacities of a creature are destined sooner or later to be developed completely and in conformity with their end.' The second proposition identified what is natural in rational humanity. He then went on to state that this rationality is realised over time, not in one individual but in the historical achievement of civil society. This deep point underlay Kant's political thought and the main purposes of his essay. What I most want to emphasise, however, is the statement which followed the first proposition about a purpose in nature: 'For if we abandon this basic principle, we are faced not with a law-governed nature, but with an aimless, random process, and the dismal reign of chance replaces

the guiding principle of reason.'[40] That is, Kant stated, we can
reason about nature in a way that produces order out of experience on the condition that we assume that there is a purpose
in nature. Similarly, we can reason about human nature as it
works itself out in history on the condition that we assume
that there is a purpose in or behind this history. If we do not
recognise this condition, we must live our lives in the midst of
'the dismal reign of chance'. With this statement, we might
think, Kant showed prevision about a fear which haunted
intellectuals two centuries later.

When he wrote on reason and purpose in history, Kant was
close to the views of his former student, J. G. Herder, the
inspired author of descriptions of the human self-creation of
language, and with language, culture, as the working out of
the immanent, God-given potential of the soul. Herder set
down an empirical, historical account of human nature, which
he understood in terms of the soul's infinitely varied expression and realisation of what is true, good and beautiful, in
cultural, political and religious life. This belief in the individual source and individual variety of expressive life became
a cornerstone of romanticism. 'Every man has the standard of
his happiness within himself: he bears about him the form, to
which he is fashioned, and in the pure sphere of which alone
he can be happy. For this purpose has Nature exhausted all
the varieties of human form on Earth.'[41]

Herder, like Vico before him, made a decisively important
argument about human distinctiveness. In his view, what distinguishes the human sphere is the human *historical* creation
of what humans themselves are. Kant predicated this self-
forming on the conditions of cognition itself. Together, these
arguments are at the root of German idealism in philosophy
and what later became a connected tradition of philosophical
anthropology. The first and crucial step was taken in the 1790s,
when J. G. Fichte reanalysed the forming of knowledge as an
act of freedom in which reflection reflects on itself. Inspired

---

40 Kant, 'Idea for a universal history', p. 42 (italics removed from
   first quote).
41 Herder, *Philosophy of the History of Mankind*, p. 78.

by Kant but, as he believed, going beyond Kant, he posited absolute freedom of what he called the 'I'. For Fichte, what is certain is that all experience presupposes, or is referred to, an 'I'. This 'I' is intuited not deduced, and we must conceive of it as the assertion of a purely free act, since anything else imaginable depends on it. This 'I' or free act posits its own being and, as this being seeks to know itself, it posits the being of something other than itself – but the conditions of this other are dependent on relation to the 'I'. It is the 'I' which makes the world. Such an argument made reflection, the act of positing being, into an ontological and not only an epistemological, condition. As surely as being 'is a self [an "I"], it must have the principle of life and consciousness solely within itself. And thus, as surely as it is a self, it must contain unconditionally and without any ground the principle of reflecting upon itself.'[42] On this basis, he then claimed to derive 'centripetal' knowledge of self as freedom and 'centrifugal' knowledge of that which self reflects on. This rendered each person's being, understood as reflective consciousness, as the being on which all else turns, and it made anthropology the science at the foundation of all philosophy.

Fichte inspired, though few people were persuaded his was the last word. Hegel, for his part, thought Fichte had not coherently represented the act of self-reflection, and he undertook systematically to elaborate a philosophy which, he claimed, reveals the act of reflection as the coming into self-knowledge of rationality (expressive of 'the world spirit'). Any brief comment must over-simplify, but, broadly speaking, his aim was to establish a true philosophy of reason's knowledge of itself in place of speculative anthropology. Drawing on the intrinsic duality of the reflective act, in which consciousness is at one and the same time consciousness of an object and of itself, Hegel elaborated an original and highly influential account of how, thereby, consciousness loses itself, so to speak, in the object, or becomes alienated, and is therefore in the position of constantly trying to reassert what it is for itself not for an other, if necessary, by destroying what is other. This is

42 Fichte, *Science of Knowledge*, p. 241.

the motor of philosophical and historical development, the dialectic of subject and object working itself out as the process of human history through action or work. 'Consciousness must act merely in order that what it is *in itself* may become explicit *for it*; in other words, action is simply the coming-to-be of Spirit as *consciousness* . . . Accordingly, an individual cannot know what he [really] is until he has made himself a reality through action.'[43] Hegel's own approach to what an individual 'is' was entirely *a priori*, and others were quick to point out that this led the sciences into a dead end.

In reaction against Hegel, therefore, philosophers attempted to build bridges between empirical description of being human as part of nature and intuitive knowledge of being human as free spiritual activity. This reaction against Hegelian idealism was, in the words of one philosopher, an 'anthropological turn'.[44] Like Kant, philosophical anthropologists sought to go beyond the opposition of determinism and freedom in order to understand the conditions of being human in which we represent events as determined or free. The main Kantian legacy, however, derived from the analysis in which Kant posited teleological propositions as logically necessary to knowledge of the phenomena of both art and life. His third critique appeared to have established grounds, and if not fully satisfactory grounds at least authority, for thinking the same form of understanding applicable to human creations and the organic creations of nature. Modern naturalistic philosophers dismiss claims about teleology in nature as a false projection of descriptions appropriate for human intentional actions into phenomena where they do not belong. (This is 'anthropomorphism'.) In contrast, Kant appeared to justify belief that the teleological form of knowledge is a condition of having certain kinds of knowledge at all, in nature as well as in human life. Thus, in German-language philosophy influenced by Kant there was not always the sharp separation between mechanistic nature and purposive mental action which was such a feature of English-language debates on 'man's place in nature'.

---

43 Hegel, *Phenomenology of Spirit*, p. 240 (§ 401).
44 Ermarth, *Dilthey*, pp. 52–4. See Toews, *Hegelianism*.

German-language writers on philosophical anthropology felt that it was legitimate to refer to 'spirit', or at least purposive agency, as a principle of action in both organic and human worlds. As a consequence, they hoped that a philosophy integrating 'man' and nature would confirm not threaten human values. Not uncommonly, as part of their rhetoric, they portrayed Goethe as having had an approach to science which unified ideals and the study of nature. Moreover, Goethe had indeed found support in Kant's third critique for belief that 'esthetic and teleological judgment illuminated one another'.[45] How securely Kant had established grounds for this kind of confidence in philosophical anthropology is another matter.

Philosophical anthropology was not collectively named as such before the twentieth century. The generation after Hegel, however, had already turned to delineate the objective and universal dimensions of value-oriented, purposeful human existence. There were university professors, like Hermann Lotze, whose book on human nature, *Mikrokosmus* (1856), was subtitled 'towards an anthropology'. But it was opponents of official university culture, like Ludwig Feuerbach and the young Marx, who most profoundly reshaped the project. Feuerbach and Marx, in revolt against Hegel and the society they thought he stood for, nevertheless took from Hegel his central idea that being human involves a historical development in which people are the subjects and objects of their own actions. In the words with which Marx distinguished people from animals: 'The animal is immediately one with its life activity. It is not distinct from that activity; it *is* that activity. Man makes his life activity itself an object of his will and consciousness.' Marx, of course, vehemently attacked the idealist presumptions of earlier anthropology and instead conceptualised being human in terms of what he thought are actual human necessities. A true anthropology must, he argued, first of all recognise that what people are is a result of the way, over historical time, they have organised work to satisfy these necessities.

45 Goethe, 'The influence of modern philosophy' (1820), in *Scientific Studies*, p. 29. See Boyle, *Goethe*, pp. 35–50; Richards, *Romantic Conception of Life*, pp. 421–57.

'It can be seen how the history of *industry* and the *objective* existence of industry as it has developed is the *open* book of the essential powers of man, man's psychology present in tangible form.'[46] This was the point where he most strongly took issue with Feuerbach. Whereas Feuerbach wanted anthropology to recognise the bodily foundation of what a person is, that is, he wanted to recognise human sensuous nature as the basis for a new moral and social order, Marx argued that we must recognise how people's bodies and minds take shape in the historical organisation of material production. Marx incisively dismissed Feuerbach, since the latter 'never arrives at the really existing active men, but stops at the abstraction "man", and gets no further than recognizing "the true, individual, corporeal man" emotionally, i.e. he knows no other "human relationships" "of man to man" than love and friendship, and even then idealized. He gives no criticism of the present conditions of life.'[47] What man has is a social nature, not some individual biological, moral or spiritual essence. 'The essence of man is no abstraction inherent in each separate individual. In its reality it is the *ensemble* (aggregate) of social relations.'[48]

In spite of his reputation as a materialist, however, Marx never restricted his view of human necessities to the merely material; he himself was one of the most educated of men. And in spite of his crucial characterisation of being human in terms of 'the ensemble of social relations' rather than human nature – though this has been subject to much debate – it appears that an ideal of what a human might really be as a freely reflecting and acting subject underlies all his judgements about political institutions. An anthropology, with the same roots as idealist anthropology, underlies his central moral and political view that the nature people express in the way they live is a historical product of their estrangement from

---

46  Marx, 'Economic and philosophical manuscripts', pp. 328, 354.
    Geras, *Marx and Human Nature*, criticised the view, once held,
    that Marx had no theory of human nature.
47  Marx and Engels, *The German Ideology*, p. 37.
48  Marx, 'Theses on Feuerbach', in *ibid.*, p. 198.

themselves by the mode of production. 'Nature as it comes into being in human history – in the act of creation of human society – is the *true* nature of man; hence nature as it comes into being through industry, though in an *estranged* form, is true *anthropological* nature.'[49] Anthropological knowledge requires an understanding of how, over historical time, the nature of people has become what it is through particular ways of organising work, rather than what it might be were there to be an objective, revolutionary consciousness of this process. All the same, Marx never systematically explained the anthropology on which his social and political thought appears to depend and this left interpretive problems when later Marxists declared he had systematically founded human science. In particular, it was argued among Marxists whether or not, or in what sense, Marx held a notion of 'species being', a Hegelian term referring to the idea that a human being has a cluster of essential powers and capacities expressed in natural and sensuous existence. (Feuerbach clearly held such a notion of species being.)

A generation later again, Wilhelm Dilthey examined at length many of the underlying issues of anthropology (though he did not particularly use this word), with much attention to historical knowledge and the life or activity of the human spirit. His work is much more representative of German academic opinion than Marx's, since he took it for granted that being human expresses something of spiritual significance, and he wrote in the idealist idiom prevalent in the academic world. Dilthey, like many other writers around 1900, sought a philosophical account of what people all too easily called 'life', by which they meant self-forming, purposive agency. Different writers found such agency in the cosmos at large, in organisms, in human history and in the achievements of special human groups or even individuals. All shared the view that mechanistic science could not understand what is central for humanity, its value-achieving accomplishments in history, the state, the arts and the sciences. They turned to what was in effect philosophical anthropology to make up for the deficiency.

49 Marx, 'Economic and philosophical manuscripts', p. 355.

This was the case, at least according to one interpretation, for even so sober an observer of contemporary society as Max Weber. In Wilhelm Hennis' view, Weber had a 'central interest' running through all his work, the anthropological theme of how it is in fact humankind's nature, through cultural and material development, to realise what has value. Weber's project was 'nothing less than the establishment of the genesis of modern men . . . via a historical-differential investigation!'[50] Throughout his multi-faceted career, in which he was finally politician as well as scientist, Weber consistently attempted to understand, and thereby to reconcile, objective knowledge of what is the case and the effort to infuse knowledge with the significance it must have if it is to be worth pursuing in the first place. His conclusion was in the tragic mode: modern people individually and modern societies at large seek, and to a degree achieve, rational order; but this is at the expense of an organic relationship between knowledge and what is thought good or beautiful – what makes order worthwhile. Weber resonantly referred to the 'disenchantment' of the modern world.[51] As is the case with Marx's notion of estrangement, it is hard to understand the powerful sense of loss which Weber's word conveyed without reference to an ideal notion of what is human. The attempt to make clear what this ideal, or ideal essence, is lay at the heart of the anthropological tradition in philosophy, actually so called, in the decades after Weber's death in 1920.

The writings to which the title of philosophical anthropology is most often attached come from the Weimar years. They involved what looks, in retrospect, to be a somewhat desperate German-language search for a renewal of the basic principles which should guide cultural life. In addition to the calamity of World War I, the authors perceived a threat from mass politics to what they believed were real values. Books by the phenomenologist Max Scheler and by Helmut Plessner, among others, voiced once again the feeling that current intellectual resources were failing to address the human condition. The

50 Hennis, 'Weber's "central question"', p. 156.
51 Weber, 'Science as a vocation', p. 30.

language of cultural crisis, it is true, went back at least into the late nineteenth century; indeed, the attempt to demonstrate that civilised culture achieves harmony between rational knowledge, practice and ideal values, and thus expresses the real purpose of human existence, had been in disarray since the attack on Hegelian thought in the 1830s. But what had most emphatically made an intellectual difference by the end of the nineteenth century was the extraordinary success of the natural sciences both in explaining the world and in securing the basis for technological progress. For all the clarity and detail which the sciences achieved in stating physical facts, this seemed, to critics, not to contribute to ethical and spiritual culture. Philosophical critics indicted the positivist stance in natural science, which limits knowledge to statements about observed material facts, for having no moral or spiritual content.

A similar indictment extended to other areas of scholarship and in particular to history. By explaining human actions and events through detailed knowledge of local circumstances, the historians, for all their protestation, appeared to be establishing a relativistic view of civilisation. How could such history contribute to a true and universal view of human purposes? By the 1930s, when political events had created a menacing backdrop to fears of intellectual failure, some voices had become tortured. Edmund Husserl, after many years of attempting to provide a new start for philosophy, phenomenology, wrote: 'But can the world, and human existence in it, truthfully have a meaning if the sciences recognize as true only what is objectively established [as matters of fact]... and if history has nothing more to teach us than that all the shapes of the spiritual world, all the conditions of life, ideals, norms upon which man relies, form and dissolve themselves like fleeting waves, that it always was and ever will be so, that again and again reason must turn into nonsense, and well-being into misery?'[52] This was the voice of anguish about the way knowledge of the facts of nature and of human life gathered by the natural and human sciences had failed to deliver knowledge meaningful for the human condition. Husserl's own answer, even though

52 Husserl, 'The crisis of the sciences', pp. 6–7.

he was old and ill, was to return again to the foundations of philosophy. By contrast, the philosophical anthropologists, some of whom – like Scheler – Husserl had deeply influenced, thought that there were ways more directly to draw together human values and knowledge about people as physical and biological beings.

Scheler, Plessner and, later, Arnold Gehlen proposed to return to an accessible description of both the facts and values of what it is to be human. They made an attempt 'to grasp the "fundamental structures of humanity" through comparison of man and animal, and through critical examination and appropriation of as many findings of the natural and cultural sciences as possible'.[53] They started with the conviction that values and purposes are observable universal characteristics of real human existence. (What they called 'real', Marxist materialists called 'ideal'.) Thus one historian of philosophy described Plessner's anthropology as 'a project for an empirical science of man with an interpretative, meaning-establishing and identity-securing purpose'.[54] The philosophical anthropologists, it is important to note, were not against science but criticised the existing human sciences for not being authentically empirical, that is, for not starting from the observed realities of meaning and significance that life has for people. For the anthropologists there is a human essence visible in the ideal achievements of cultural life, like great works of art. While not at all anti-science, they did indict mechanistic science. Plessner, for example, exploited contemporary descriptions of organisms as having organic relation with their life-situation, and he also pictured the human pursuit of values as embedded in nature. At the same time, he made it plain that he thought it is precisely the human achievement of value-expressing life that makes humankind distinct. Such views, it is worth observing, were not restricted to German culture at this time. The eminent English neurophysiologist C. S. Sherrington, for example, wrote similarly in *Man on His Nature* (1940). But Sherrington chose a poetic rather than a philosophical voice in which to

---

53 Honneth and Joas, *Social Action and Human Nature*, p. 41.
54 Schnädelbach, *Philosophy in Germany*, p. 224.

express his views and, as befitted a meticulous experimental scientist, he was much more circumspect about restricting what he claimed to know to established facts.

The anthropological interest in 'the structures of humanity' at times took a direction radically detached from the natural sciences. This was true for the philosophers whose starting point, like Fichte's, was the irreducibility of human freedom, and whom the United States scholar Walter Kaufmann linked together as the existentialist tradition. Heidegger's work, *Sein und Zeit* (1927), gained a reputation as a new basis for an authentic anthropology based on the re-encounter with existence as it is, rather than as philosophical traditions have come to portray it, though Heidegger made plain that he intended his book to lay foundations but not carry out an anthropological project.[55] Some writers, like the philosopher José Ortega y Gasset, even took a radically anti-naturalistic stance, distancing themselves from any possible account of human nature. Ortega wrote: 'For man has no nature . . . Man is no thing, but a drama – his life, a pure and universal happening which happens to each one of us and in which each one in his turn is nothing but happening.'[56] This was idealism of the purest kind, distancing itself from philosophical anthropology and biology alike.

Ernst Cassirer was the philosopher who perhaps came closest to building a bridge between the world of philosophical anthropology, with its roots in Kant and German idealism, and Anglo-American empiricism. By dint of his vast philosophical and historical range, his deep knowledge of the modern natural sciences and his forced emigration from Germany, through Britain and Sweden, to the United States, Cassirer attracted attention as a philosopher who might indeed make possible a humanistic science of man, to use a then current expression.

Cassirer argued, writing in German in the 1920s, that the mode of shaping knowledge about humankind must be appropriate for a subject matter characterised by its ability to express symbolic form. He took a position like Kant's in the

55 Heidegger, *Being and Time*, p. 38.
56 Ortega, 'History as a system', p. 303.

theory of knowledge (he was an eminent Kant scholar), and
analysed knowledge in terms of what he called 'form', the act
in which reason structures experience. This led him, in effect,
to propose a philosophical anthropology in which he answered
the question 'What is human?' in terms of the human capacity
to create a symbolic world: 'Man's outstanding characteristic,
his distinguishing mark, is not his metaphysical or physical
nature – but his work. It is this work, it is the system of human
activities, which defines and determines the circle of "humanity."' This work has symbolic form in language, the sciences,
the arts and human institutions – in short, in culture. Man is
'an *animal symbolicum*'.[57] Cassirer was well informed about
evolutionary biology, if in the contemporary German terms
which genetics and molecular biology elsewhere discarded,
and he fully rejected conservative arguments that some kind
of unique entity separates human beings from nature. But he
did hold that there is a unique human capacity to construct
different symbol systems as different 'instruments' for living
in the world. What distinguishes the human sphere, he argued,
'cannot be understood and described by pointing out specific
distinguishing features. For the decisive change lies, not in
the emergence of new features and properties, but in the characteristic *change of function* which all determinations undergo
as soon as we pass from the animal world to the human
world.'[58] Humans, like animals, act instrumentally in the world,
but reflective awareness makes possible different symbol systems – in the sciences, in the arts, in morality – for different
purposes. By this route, Cassirer arrived at a position which
was attractive to English-speaking humanists: it took seriously
both the explanatory and instrumental power of modern natural science, as more extreme idealists (like Ortega) did not
do, and yet it also represented that power as the forming
activity of the human spirit.

Cassirer also wrote scholarly intellectual history. Historical
work, he argued, is both the route to empirical knowledge of
human symbolic activity and the means to analyse logically

57 Cassirer, *Essay on Man*, pp. 68, 26.
58 Cassirer, *Logic of the Humanities*, p. 73.

the rational conditions which make possible its 'formation'. '"*Humanitas*," in the widest sense of the word, denotes that completely universal – and, in this very universality, unique – medium in which "form," as such, comes into being and in which it can develop and flourish.' A full notion of humanity encompasses the creation of 'form' in all its variety – in the physical sciences, in the social and cultural sciences and in the arts, as well as in the creation of civilised life. While he certainly thought it necessary to integrate anthropology with biology, he did not think it possible to provide a causal explanation of 'forming' activity; as he indicated, it simply *is* the being of being human. As such it distinguishes the sphere of what is human. 'It is this process which distinguishes the mere *transformation* [*Umbildung*] taking place in the sphere of organic emergence from the *formation* [*Bildung*] of humanity.'[59]

If Cassirer did not approach the human work of imposing 'form' as a phenomenon to be explained in causal evolutionary terms, he was nevertheless interested in how to represent human capacities as the evolutionary emergence of new functions for existing structures. He paid particular attention to the work of the Baltic German biologist Jakob von Uexküll, who had introduced the notion of an animal's, or human's, life-world (*Umwelt*) as a way of describing organism and environment as a unified system. Cassirer conceived of concepts as 'instruments' in the human life-world. 'All theoretical concepts bear within themselves the character of "instruments." In the final analysis they are nothing other than tools, which we have fashioned for the solution of specific tasks and which must be continually refashioned.'[60] In this respect, Cassirer came close to the ideas of the United States evolutionary philosopher, John Dewey, who also, while not explaining consciousness in causal terms, conceptualised it as the peculiarly human instrument of evolutionary adaptation.[61]

The post-1945 years inevitably saw much harsher attitudes in European philosophy; the days when it was possible

---

59 *Ibid.*, pp. 22, 216; see also pp. 176–81.
60 *Ibid.*, p. 76.
61 Clarence Smith Howe, introduction to *ibid.*, p. xi.

to refer unblushingly to the human spirit were gone. Critics thought the project of philosophical anthropology terminally compromised by the actual course of events: *this*, one might say, is what humankind is. The longing to reassert so-called real values appeared to be nostalgia for the past of European highbrow culture and for the authoritative position of the most educated class within it. The rhetoric of true values, grasping meaning and the pursuit of the essence of man, which the historian Fritz Ringer tied to a 'mandarin' academic world, had always been vulnerable to the accusation of vacuity.[62] The philosopher of culture, Heinrich Rickert, for example, had referred to the 'objectively "spiritual" content' of culture – 'the sum total of that which is not perceptible by the senses, but which can be grasped only in a nonsensorial manner and which gives life its importance and meaning'.[63] It is perhaps this kind of statement that led Geertz to note, implicitly contrasting explanation in the natural sciences and in the humanities, that 'the besetting sin of interpretive approaches to anything . . . is that they tend to resist, or to be permitted to resist, conceptual articulation and thus to escape systematic modes of assessment'.[64] Even more seriously, the preoccupation with the nebulous ideals characteristic of philosophical anthropology had distanced intellectual work from the specific social and political conditions of contemporary life, and thereby, to a degree, Ringer argued, contributed to the circumstances in which the Nazis gained power.

All the same, the feeling that life should have an 'importance and meaning', as Rickert had emphasised, and that science was somehow lacking, or inadequate, as a way of addressing this, did not go away. The questions at the heart of Kant's attempt to bridge necessity and freedom, Dilthey's concern with knowledge adequate for 'life' and Husserl's troubled awareness that positivist knowledge leaves values completely unanchored all remained. But new terms and new thoughts had to be found.

---

62 Ringer, *Decline of the German Mandarins*.
63 Rickert, *Science and History*, p. xv.
64 Geertz, 'Thick description', p. 24.

Writings on philosophical anthropology, albeit in new dress, did continue to appear after 1945. Jung's psychological anthropology, for example, with its rich descriptions of what he claimed to be the universal unconscious structure of the human soul, attracted a huge audience. Critics, however, thought the whole tradition of anthropology tainted by an irresponsible imprecision about the actual material and social source and nature of values. A number of these critics, though, took inspiration from Marx's early writings, writings which were themselves exemplary of the 'anthropological turn' in German philosophy after Hegel's death. There was also the work of the Frankfurt school to draw on, dating from the Weimar years and from the period of the school's exile in the United States. Social philosophers who kept the word 'anthropology' at a distance because of its idealist connotations nevertheless, in their moral and political critique of the social and economic conditions of life, returned to themes that the anthropologists had also taken up. They explored the philosophy of being human as a bridge between biological knowledge and social understanding relevant to political action. The German social philosophers Axel Honneth and Hans Joas wrote: 'Neither can a subdiscipline of biology serve to establish what enters into the central categories of the social and cultural sciences as the fundamental characters of the human being; nor can philosophy immediately grasp cognitively the "essence" of the human being by means of abstract definitions.'[65] What is needed, they believed, is an understanding of the way historical culture mediates biological nature in the life of actual people and opens up the human world to ideal goals. This had also been Marx's project.

The work of Jürgen Habermas in the 1960s became a focus for many of these issues. His notion of human 'knowledge-constitutive interests' then appeared to be a constructive way forward for discussing the relationship between ideal goals and historical people. His work was grounded in post-Kantian German philosophy and well aware that it was a renewed attempt to address anthropological themes, especially the

65 Honneth and Joas, *Social Action and Human Nature*, p. 7.

relationship between biology and culture. Habermas found expressions which tied together the species' 'natural history', the historical self-formation of what is human in the particularities of culture and the practical implementation of knowledge in ways of life. This made possible a theory of knowledge in the different sciences, natural and human, as forms of *praxis*, the human activity in which the natural sciences develop through interaction with physical nature and the human sciences develop through social relationships.

Habermas posited different 'knowledge-constitutive interests', rather as Cassirer had posited the more idealist conception of 'forms', in his theory of knowledge, and Habermas was as emphatic as Cassirer that though these interests 'mediate the natural history of the human species with the logic of its self-formative process . . . they cannot be employed to reduce this logic to any sort of natural basis'.[66] As this suggests, a tension, evident earlier in anthropology, remained in Habermas's work, between an essentialist strand of argument concerned with what in logic must be the conditions of all reason and a pragmatic strand recognising the contingent historical circumstances of specific reasoning. Where Habermas did diverge from Cassirer was in his attention to social theory and the classification, in the manner of Weber, of what different human 'interests' actually are. He distinguished three and linked them to the different fields of science: 'The approach of the empirical-analytic sciences incorporates a *technical* cognitive interest; that of the historical-hermeneutic sciences incorporates a *practical* one; and the approach of critically oriented sciences incorporates the *emancipatory* cognitive interest.'[67] This classification, dividing knowledge into the instrumental, making the world an object for technology, the practical, making the human world subject to social organisation, and the emancipatory, making politics open to reason reflecting on itself, had plausibility. But readers in later decades became sceptical about the attempt formally to ground a classification of knowledge

66 Habermas, *Knowledge and Human Interests*, p. 196.
67 Habermas, 'Knowledge and human interests' (Inaugural Address, 1965), in *ibid.*, p. 308.

in critical reason. The successive failure of philosophers from Kant to Habermas (in this early work) to establish definitive and authoritative grounds for the theory of knowledge strongly suggested – though in itself hardly proved – that there was something wrong with the search. This was Foucault's conclusion, as, in a different way, it was that of the United States pragmatist philosophers and of Richard Rorty. This seemed to mark the end of philosophical anthropology as German-language scholars had understood it.

## Philosophical anthropology scorned

Dilthey's writing exemplifies a humanistic anthropology, a view of the human condition which starts out from belief in the autonomous dignity of each person resulting from free will and reason. 'A man finds in . . . [his] self-consciousness a sovereignty of will . . . a capacity for subordinating everything to thought . . . [and] by these things he distinguishes himself from all of nature.'[68] Educated, reflective culture in Europe and North America, from the Enlightenment to the mid-twentieth century, aspired through art, science and social and political action to build a world commensurate with such belief. The ideal, however, always had its critics and, even more, its sceptics. In the twentieth century, many concluded that statements like Dilthey's were fantasy, idealistic as opposed to realistic, and remote from the experience and needs of the mass of the people. The catastrophes of the century brutally mocked ideals. Even before events withered humanism, Nietzsche had subjected the very notion of reason, as it had been previously understood, to a blistering critique. Yet he held out the hope of transforming reason to meet the contingency of its own inadequacy. In contrast, Dostoevsky had indicted Enlightenment reason in general as a base for morality and imagined a return to Christian truths as embedded in native Russian tradition.

A century later, Foucault became a powerful and influential voice critical of philosophical anthropology and its humanistic

68 Dilthey, *Introduction to the Human Sciences*, p. 79.

presumptions. In the publications which initially brought
him fame, Foucault directed criticism at the philosophical
phenomenology reigning in France in the 1950s. As it happens,
his unpublished secondary thesis for his higher degree con-
sisted of a French translation and commentary on Kant's lectures on anthropology, where Kant discussed the conditions
and nature of experience and conduct in ordinary life.[69]
(Foucault published his principal thesis as *Folie et déraison*,
1961.) His decision to comment on this text, rather than on
one of the great critiques, was perhaps itself a signal of scepticism about the possibilities for the kind of 'transcendental'
rationality which Kant, and later the phenomenologists, had
sought. Subsequently, in *Les Mots et les choses* (1966), scepticism turned to antagonism, and Foucault made plain that he
regarded the large claims of philosophical anthropology as fit
only for scorn. He attacked anthropology and its associated
humanism: the subject, 'man', is neither the historically timeless nor the unified being which anthropologists claim.
Foucault argued that we should resist the 'blackmail' of
Enlightenment ideals: we do not have to be for or against reason but may instead understand how reason constitutes 'man'.
He proposed 'a historical investigation into the events that
have led us to constitute ourselves and to recognize ourselves
as subjects of what we are doing, thinking, saying'.[70]

Philosophical anthropologists, and in this respect Dilthey is
again exemplary, appeared in something of an impasse, committed at one and the same time to belief in a human essence
(reason or freedom) and to the historical particularity of actual
people and their actions. For French intellectuals the dilemma
was embedded in the problematic relations between phenomenology and Marxism. As a result, 'the unstable tensions
between a theory of man based on human nature and a dialectical theory in which man's essence is historical lead to a
search for a new analytic of the subject. One sought a discipline
which both has empirical content and yet is transcendental,

69 Eribon, *Foucault*, pp. 63, 90, 110–15.
70 Foucault, 'What is enlightenment?', p. 46. See Foucault, 'Nietzsche,
    genealogy, history'.

a *concrete a priori*, which could give an account of man as a self-producing source of perception, culture, and history.'[71] The tensions precipitated a re-examination of the knowing subject, the Cartesian 'ego'. Foucault's step, in the 1960s, was to conceive of this subject, 'man', as one which became an object of knowledge only in the modern age. This was a very surprising and, to many English-language readers, incomprehensible view.

Foucault's argument, in the concluding two chapters of *Les Mots et les choses*, is difficult to assess; moreover, Foucault later distanced himself from what he said there. Almost everyone rejects the claim, which he then made, that 'man', the subject of the human sciences, was invented around 1800.[72] In summary, he argued that, at this time, a radical shift in the articulation of knowledge created biology, economics and philology – for him, the three sites or models of disciplines in the human sciences – replacing discourses of classifying, exchanging and speaking by discourses of life, labour and language. During most of the eighteenth century, he maintained, the conditions of knowledge made knowledge appear the transparent representation of the world, human nature included, but did not allow for 'man' as reflexive being. The conditions of modern knowledge, in contrast, he held, led to the replacement of the notion of transparency of representation by the notion of 'man' as the subject of life, labour and language. In life, labour and language, that is, in the human sciences which claimed knowledge of these activities, a subject appeared which represented itself to itself. Discussing the reflexivity of language, Foucault wrote: 'The object of the human sciences is not language (though it is spoken by men alone); it is that being which, from the interior of the language by which he is surrounded, represents to himself, by speaking, the sense of the words or propositions he utters, and finally provides himself with a representation of language itself.'[73]

71 Dreyfus and Rabinow, *Foucault*, p. 33.
72 For example, Wittrock, Heilbron and Magnusson, 'Rise of the social sciences', p. 7; Wokler, 'Enlightenment and the French revolutionary birth pangs', p. 61.
73 Foucault, *Order of Things*, p. 353.

Foucault thus argued that the human sciences came into existence, and could only have come into existence, in modern times. Only since the late eighteenth century has there been the kind of reflection in which people make themselves both subject and object of study, and which, systematically pursued, forms the human sciences. After Kant, we face the knowledge that we cannot know the world 'in itself' but necessarily know through the framework given in the actual exercise of reason. This 'finitude', Foucault argued, is the condition that has turned reason to reflect on itself and in this reflection to create the domain in which the modern human sciences establish facts about what it is to be a human being. 'Man appears in his ambiguous position as an object of knowledge and as a subject that knows . . . Man's finitude is heralded – and imperiously so – in the positivity of knowledge: we know that man is finite, as we know the anatomy of the brain, the mechanics of production costs, or the system of Indo-European conjugation.' As a result, he continued, 'our culture crossed the threshold beyond which we recognize our modernity when finitude was conceived in an interminable cross-reference with itself'.[74] Modern understanding of what humans are – and there is no truth beyond this understanding – is limited to the self-referential terms of a historically bounded way of thought (or *discours*).

This destabilised the notion of the knowing 'I' which had been at the heart of anthropology and the human sciences since Kant, and which was the centre of the existentialist account of freedom current when Foucault formulated his criticisms. When Kant himself taught empirical psychology (in the eighteenth-century sense), his starting point was just such a notion: 'The first [condition for experience] is the consciousness of myself, the I, it is the first act <*actus*> of the mind <*psyche*>: the faculty for cognizing oneself as representing subject, and also as object of our own representation.'[75] New arguments in the 1960s, among them Foucault's, relocated this reflective capacity away from a 'self' and into language,

74 *Ibid.*, pp. 312–13, 318.
75 Kant, *Lectures on Metaphysics*, p. 372.

with clear and deliberate anti-humanist implications. Foucault, for example, without mentioning his name, absolutely opposed the humanistic work of Gusdorf. Gusdorf was then authoring a multi-volume history of the human sciences, from ancient to modern times, written specifically for the humanistic reason that these sciences 'doivent être la mauvaise conscience des sciences sans l'homme, des sciences inhumaines'.[76] Without the human sciences, in Gusdorf's view, science is 'inhuman'. According to Foucault, such projects were based on false premises and took for granted an ontology of the human spirit or the self or human essence. It was, Foucault argued, precisely the capacity of language to denote 'the self' or 'the human', and the activity of the sciences in making 'the self' and 'the human' the object of truth claims, that should be the subject matter of a philosophical human science. (Hence the 'scare quotes' around every term.) For Foucault, therefore, philosophical anthropology is a historically contingent, and passing, intellectual tradition, and its claim to reveal a real, shared humanity, not worth discussion. In subsequent work, he concentrated on the way everyday practices of discipline and government constitute knowledge in the human sciences and thereby embed the modern manner of being human. He attended to the way the reflexive circle given in human finitude recirculates power in 'regimes of truth' rather than to what he thought dead issues of anthropology.[77]

Unsurprisingly, all this was most controversial, and this and the brilliant and original writing ensured a large audience. Readers linked Foucault's work with the claim, especially associated with Derrida, that knowledge cannot, so to speak, escape the language which expresses it and be referred to something beyond language. Foucault elaborated an influential view of what this meant for the human subject. To humanist critics, this substituted a formalistic anti-humanism for a living commitment to the unique qualities of being human. But in answer to such critics it could be said that

76 Gusdorf, *Introduction aux sciences humaines*, p. 511. See Gusdorf, *Les Sciences humaines* in twelve volumes.
77 See Foucault, 'Truth and power'.

Foucault had only drawn to the surface, and made inescapable, longstanding questions which wishful thought about 'the
human' as an ideal had refused to confront. I return to these
questions under the heading of reflexivity.

There have, of course, been attempts to establish a critically
aware middle ground. Honneth and Joas, for example, while
recognising that Dilthey's anthropology was untenable, found
Foucault's stance self-defeating – since Foucault himself, in
critically revealing the limits of reason, deployed reason's own
terms. As an alternative, they argued, 'anthropology must not
be understood as the theory of constants of human cultures
persisting through history, or of an inalienable substance of
human nature, but rather as an enquiry into the unchanging
preconditions of human changeableness'.[78] Foucault would
surely have thought that this let in through the back door the
belief in 'unchanging preconditions' which had just been
pushed out through the front. All the same, Honneth and Joas
had a point. The actual practice of writing philosophy, or of
engaging in reflection on the human sciences, itself presupposes certain principles of reasoning, however much it is the
purpose of writing to show that principles of reasoning originate in historically constituted practice. Foucault was intensely
aware of this, which partly explains his use of a rhetoric which
destabilises its own presumptions. Moreover, in his political
activity which began in the late 1960s, Foucault in practice juxtaposed different representations of what it is to be
human, and judged between them, and this appears to allow
that there could be an objective account of representation as
an act in itself. What I would stress is that these were not just
Foucault's problems but everyone's. By exposing them, he made
his own writing central to the subject matter of the human
sciences.

Honneth and Joas had a vision of anthropology as the
humanisation of nature, an integration of knowledge in natural science with moral knowledge of human culture. Their
programme exemplifies the attempt to construct a social science in the service of humane values. 'First, the human being

78 Honneth and Joas, *Social Action and Human Nature*, p. 7.

humanises nature; that is, he transforms it into what is life-serving for himself and thereby creates, in an interknitting of the transformation of nature and the development of the human personality . . . Second, the human being humanises nature within himself in the course of the long civilising process . . . Lastly, the human being himself is a humanisation of nature . . . in the human being, nature becomes humane.'[79] This presupposes an ontology of being human, a base line, taking the form of a moral claim about what it is to be human and humane. By contrast, the proponents of an evolutionary synthesis of knowledge of human nature made this base line a claim about what is biologically real in the roots of the human condition. In further contrast, religious writers addressed the base line as soul or spirit. Foucault set out to analyse the conditions in reason for all such claims and thereby to describe how the finitude of the human condition expresses itself in reason, language and power.

It seems clear at the beginning of the twenty-first century that the challenges of philosophical anthropology cannot be met, at least not in the terms earlier proponents hoped. Moreover, there is not much future for philosophy if it does little to comprehend the social and technological transformation of human identities actually going on and instead waxes nostalgic about 'real values'. Perhaps it is not surprising that many people appear to take literally the word of evolutionary science, or the word of one or another God, and leap to beliefs about human nature for which they claim absolute objectivity. There are, however, constructive alternatives, and they stem from understanding reflexive knowledge.

79 *Ibid.*, pp. 9–10.

# 2

# Reflexive knowledge

## Reflexivity in modern philosophy

The quest for the essentially human has produced conflicting and incompatible claims. Foucault opposed all such talk about the essence of being human and, followed by many scholars in the social sciences and humanities, turned to the manner in which different discourses construct the subject, 'man'. Independently of philosophy, the sheer number and diversity of disciplines with some claim to a place in the human sciences has led to a feeling of hopelessness about searching for a unified answer to Kant's question, 'What is man?'

There is, however, a philosophical point, or rather cluster of related points, which enables us to view the situation as the outcome of fundamental conditions of knowledge not as a stalled, or even failed, quest. I shall call this 'reflexivity'. Many other authors have used the term in relation to philosophical matters of great range and depth, which I will not exhaust.[1] I use the word to refer both to the examination of the unfounded assumptions in *any* body of knowledge, and to the process whereby knowledge of what is human changes what it is to be human. The discussion here prepares the ground for the claim that history is knowledge of what it is to be human and that history, especially the history of the human sciences, is a human science. Only if we work through the reflexive dimensions of knowledge, I suggest, will we find a way to move beyond

---

1 See Smith (ed.), 'Reflexivity'. For reflexive notions of 'the self', which I do not discuss: Seigel, *Idea of the Self*, pp. 17–32; Taylor, *Sources of the Self*, pp. 130–2.

the conflicts between philosophical anthropology, biology and other approaches to human nature.

Cassirer began his anthropological *Essay on Man* (1944) with the statement 'that self-knowledge is the highest aim of philosophical inquiry appears to be generally acknowledged'.[2] It is indeed commonplace to assume that the capacity for self-reflection, the capacity to turn conscious awareness into an awareness of itself as an activity, is central to what defines a person as a person. Descartes's description of the individual human mind as thinking being, the 'I', stands – in the conventional picture – at the beginning of modern discussion of reflection. Subsequently, one strand of opinion compared so-called external observation, via the five senses, and internal observation, in which the mind turns its knowing activity on itself. Locke called the latter activity reflection and referred to 'the *perception of the operations of our own minds* within us'.[3] Nearly a century later, Kant, in his anthropological writings, stated: 'Concepts are not innate in us, but rather the capacity to reflect.'[4] He then went on to analyse the logical form which reason must have if the kind of reflection we do in fact undertake does indeed make coherent knowledge possible. This led directly to the work of the German idealists, who shaped interest in reflection as a matter of making reason transparent to itself and of making clear the logical and ontological conditions for reasoning to occur. From this turning of reflection into the agent of reflection on itself come modern discussions.

English-language writers, following Locke, focused on the formation of knowledge through sensory experience. This way of thought tended to liken the mind to a mirror reflecting the world, and hence it was natural to imagine self-knowledge as also a kind of reflection. Philosophers took reason to be like a mirror, and this metaphor has persisted in theories of knowledge down to the present.[5] There were difficulties from the

---

2 Cassirer, *Essay on Man*, p. 1.
3 Locke, *Essay Concerning Human Understanding*, vol. 1, p. 78. See Taylor, *Sources of the Self*, pp. 173–6.
4 Quoted in Zammito, *Kant, Herder, and the Birth of Anthropology*, p. 275.
5 Rorty, *Philosophy and the Mirror of Nature*, p. 12.

beginning, and one, which troubled Locke's first readers and which Hume took up in a now famous discussion, concerned the very notion of the self. If knowledge comes through reflecting the world, it seems that what we call 'the self' can be only a name for the apparent continuity of sensory experience. Surely, there is nothing to observe by looking inwards other than the character and sequence of sensations? A second difficulty, which came to haunt psychology, was that it appeared very questionable to think that a mind reflecting on itself could claim objective knowledge of the kind claimed about the physical world. It was patent to many scientists and philosophers, like Comte, that it could not, and they therefore rejected reflection as a source of knowledge. The rigorously empiricist epistemologies which followed, however, proved a failure, and it would seem that some account of reflective activity must form part of any theory of knowledge.

If this is accepted, it is necessary to come to terms with arguments first made with some degree of clarity by Herder in the late eighteenth century. Our language about the world is not simply a description of the world; language actually brings into being a particular version of both the world and ourselves. Taylor called this theory of language and knowledge expressivist: 'Words do not just refer, they are also precipitates of an activity in which the human form of consciousness comes to be. So they not only describe a world, they also express a mode of consciousness, in the double sense . . . that is, they realize it, and they make determinate what mode it is.'[6] Philosophy subsequent to Herder, or at least that part of philosophy which flourished without subordination to the empirical sciences, placed this world-constituting and self-constituting activity at its centre.

The central problem with the empiricist belief that scientific statements correspond to or represent reality is that it is impossible to make any statement, even the simplest observation statement, which does not in itself contain a theoretical or conceptual presupposition. All observation statements are theory-laden. As Hegel observed, 'truth is not a minted coin

6 Taylor, *Hegel*, p. 19.

that can be given and pocketed ready-made'.[7] Conceptual presuppositions inhere in the very language used: there is no neutral or, so to speak, natural language. 'Signs do not intrinsically correspond to objects, independently of how those signs are employed and by whom. But a sign that is actually employed in a particular way by a particular community of users can correspond to particular objects *within the conceptual scheme of those users*. "Objects" do not exist independently of conceptual schemes. We cut up the world into objects when we introduce one or another scheme of description.'[8] We cannot, in the last resort, isolate a description from the activity of rendering the description. 'We cannot give a sense to this opposition between the nature of reality and the conditions of our knowledge of it.'[9] We cannot step outside our statements to find independent grounds on which to stand to produce a description which corresponds to its object independently of assumptions embedded in the prior choice of this particular description. These prior assumptions, including those that go into any act of reflection, can themselves become the focus of reflexive attention. And so on. Endless reflection on reflection appears possible.

A common-sense realism, of a kind that many natural scientists and historians as well as ordinary people implicitly adhere to, which asserts that empirical reflection gives access to what is real, and no more need be said, simply will not do. It puts the question of what reflection is into a box and demands that it be kept locked.

German-language philosophers use the verb 'to reflect' (*'reflektieren'*) to refer to an active, constructive process, in no sense like a mirror, which constitutes both mind and its objects in one way rather than another. One translator, seeking appropriate expression in English, put the point this way: 'In German usage . . . the word "reflect" expresses the idea that the act in which the subject reflects on something is one in which the object of reflection itself recurves or bends back in a way that

7 Hegel, *Phenomenology of Spirit*, p. 22 (§ 39).
8 Putnam, *Reason, Truth and History*, p. 52.
9 Hampshire, *Thought and Action*, p. 13.

reveals its true nature. The process through which consciousness reflects back upon itself, insofar as it reveals the constitution of consciousness and its objects, also dissolves the
naive or dogmatic view of objects; thus they themselves are
re-flected through consciousness.'[10] According to this view, a
mind reflecting on itself is bringing into being what a mind is.
It is not simply a subject reflecting on an object.

Hegel, when he substituted phenomenology for epistemology, started from the apparent double or reflective nature of
consciousness – that consciousness appears to involve reflection, in one and the same act, on both an object and itself.
'For consciousness is, on the one hand, consciousness of the
object, and on the other, consciousness of itself; consciousness of what for it is the True, and consciousness of its
knowledge of the truth.' And, as he went on to argue, 'Since
both are *for* the same consciousness, this consciousness is
itself their comparison'; that is, consciousness intrinsically
presents a relationship between two expressions. From this,
comparison creates a '*dialectical* movement which consciousness exercises on itself and which affects both its knowledge
and its object, [and] is precisely what is called *experience*
[*Erfahrung*]'.[11] In 'dialectical movement' both knowledge and
the object of knowledge change; this is the historical process
by which 'the Absolute' comes to its full realisation. Hegel
thus thought that he grounded reflexive activity in an ontology,
a rational account of being. But a number of later philosophers, from Nietzsche to Derrida, were to argue that it is
always possible to renew the reflective process: whatever ground
we chose to rest on can itself be made the object of reflection.
This is a process which undermines the claim of any ground
to be ultimate or absolute. Discussion therefore moved from
the topic of reflection to the topic of reflexivity.

We may compare reflection reflecting on itself to the situation when a child continuously asks 'Why?', whatever the
answer. At some point, the only possible response is to say 'It

10 Jeremy J. Shapiro, translator's note in Habermas, *Knowledge and
   Human Interests*, pp. 319–20.
11 Hegel, *Phenomenology of Spirit*, pp. 54, 55 (§§ 85, 86).

just is so'. At this point the adult presupposes something to be true which is not verified in the discussion itself (and, indeed, often remains unclear to the child). Following reason, and following the desperate parent, we stop at this point because there are limits to which we can go without becoming unintelligible and without intolerable burden. It is always possible to reject these limits, however, as the artistic avant-garde has made it its business to do, seeking to replace reason by poetics.

There is no occasion on which, given sufficient ingenuity, we cannot expose unverified or circular arguments at the base of someone's thought. Self-confirming arguments are characteristic of all cognition.[12] Any general claim about the world presupposes that the claim also applies to the claim being made. There is, in Thomas Nagel's memorable phrase, no 'view from nowhere', no point outside the circle of knowledge and its presuppositions.[13] (Unless, perhaps, if I may state it thus briefly, we think we know *by faith* that there is such a point.) Philosophers like Nagel, in the analytic tradition, thus arrived at one of the principal insights of hermeneutics, the so-called hermeneutic circle. Hermeneutics is like a dialogue, in which the person seeking to understand tests out knowledge by exposing it to an interlocutor (or text). This presupposes that the investigator has some prior knowledge about what the statements under study means, as Heidegger indicated in his remark that 'whenever we encounter anything, the world has already been previously discovered'.[14] Knowledge requires some knowledge of the meaning attached to the whole of a speech, and no part, or fact, has meaning without it; conversely without parts, or facts, no whole has content. As a result, in Gadamer's words, 'fundamentally, understanding is always a movement in this kind of circle' between parts and whole. We can make no meaningful statements about a 'world in itself'. The world 'is always a human – i.e., verbally constituted – world that presents itself to us . . . The criterion for the continuing expansion of our own world picture is not

12 See Barbara Herrnstein Smith, *Belief and Resistance*, pp. 37–51.
13 Nagel, *The View from Nowhere*.
14 Heidegger, *Being and Time*, p. 114.

given by a "world in itself" that lies beyond all language. Rather, the infinite perfectibility of the human experience of the world means that, whatever language we use, we never succeed in seeing anything but an ever more extended aspect, a "view" of the world.'[15] There can be no definitive, uniquely validated language for describing the world, as the positivists, who attempted to limit knowledge to statements expressible as observation claims, had hoped. Knowledge of the way particular uses of language form the 'view' (in Gadamer's word) of how particular people, scientists obviously included, have constituted the world, has a necessary place in knowledge. Knowledge which does not include reflection on the sources of its own language is incomplete. But whatever reflection there is on these sources, there remain presuppositions which in turn can become the subject of reflection.

Carrying on reflection on reflection is a theoretical possibility. In practice, all areas of scholarship get on with what they do while making presuppositions which in a sense beg the question, and which, while not verified, are not doubted either. As Wittgenstein observed: 'It belongs to the logic of our scientific investigations that certain things are *in deed* not doubted. ... If I want the door to turn, the hinges must stay put.' The child's parent may take comfort: 'My *life* consists in my being content to accept many things.'[16]

Natural and social scientists, historians and scholars in the humanities all live in disciplinary worlds which require research and writing within a socially given framework. This framework appears highly appropriate, even natural, for the purposes for which the discipline that adopts it exists. In these circumstances, few people have sympathy with continuous examination of what the given framework actually is, though this sympathy may grow in times of intellectual or material crisis. The potential for questioning, however, is always there, and this has been the source of much recent philosophical interest. Indeed, it could be said that there is now a discipline,

15  Gadamer, *Truth and Method*, pp. 190, 447. Also Gadamer, 'Universality of the hermeneutical problem'.
16  Wittgenstein, *On Certainty*, p. 44ᵉ (§§ 342–4).

in which Derrida was the master, of practising reflexivity, reinterpreting the presuppositions of thought as systems of self-reference in language itself. This fostered the reflexive use of language, turning language to examine itself and showing that language 'is about' language. It also assigned to language fundamental status 'as a form of being that rises above, or "cancels and preserves" (*aufgehoben*), any merely "human", sensuously grounded, social and historical form of experience'.[17] Writing in this vein has inevitably recognised that its own claims are no more grounded in something beyond language than any other writing. There is no closure of meaning, that is, no way, finally, to legislate that a statement means this and only this; there are always further possibilities of meaning. As there appears to be no further step to make in logic, some intellectuals have turned to an aesthetic rendition of the human condition, an ironic stance and self-justifying play with language.[18]

Modern intellectual life therefore appears to present stark alternatives. The first is to write without reflexivity, to get on with the discipline in hand, as most natural and social scientists do, and the second is to engage the poetics of irony, asserting and reflexively questioning the assertion at one and the same time. The gulf between these forms of writing, it scarcely needs saying, is indeed a gulf and not the least of the divisions between the natural sciences and the humanities. There is in addition, however, another side to these reflexive questions, the material culture of modern – or, as it may be, postmodern – life, and especially its technological transformation.

## Reflexivity in modern life

In early western thought there were no notions of 'pure' and 'applied' knowledge, though of course there were divisions within the classification of knowledge between a theoretical science like mathematics and a practical science like politics.[19]

---

17 McDonald, 'Language and being', p. 193; see Lawson, *Reflexivity*.
18 See Megill, *Prophets of Extremity*.
19 Gadamer, 'Hermeneutics as practical philosophy'.

Only in the nineteenth century, it would seem, was the pure/
applied distinction to become current. More recently, divisions between pure and applied science once again do not
have much purchase, in part because the funding regime for
modern science makes it so obvious to scientists themselves
that science has material purposes. Historians and sociologists of technology, in this like Marxist-influenced writers
earlier, no longer look at material history as the application of
knowledge; rather, they describe science and technology as a
set of integrated social practices.[20] The influential approach of
Michel Callon and Bruno Latour, for example, treated knowing subjects – people – and material objects, both natural and
artificial, equally and symmetrically as 'actors' within a network, the resulting form of which depends on the distribution
of power.[21] All the elements, human and non-human, have the
potential to take new shape as the balance of power changes.
These sociologists talked about actor networks and not about
reflexivity but, equally with philosophers who have subscribed
to the reflexive turn, they destabilised all claims about there
being a 'nature' independently of our knowledge or action.
They also destabilised the separation of people and their
technologies. 'To conceive of humanity and technology as
polar opposites is, in effect, to wish away humanity: we are
socio-technical animals, and each human interaction is socio-
technical. We are never limited to social ties. We are never
faced only with objects.'[22]

The language which describes people and things symmetrically is unfamiliar, but the conclusions are not at all esoteric.
Technology, visibly and vividly, is transforming what it is to
be human. New reproductive technologies, biotechnology,
genetic engineering, mind-altering drugs, information technology and transplant and reconstructive surgery are literally
reshaping, or will soon reshape, human life-forms. As a result
we frequently no longer know where to draw the boundary

20  Bijker, Hughes and Pinch (eds), *Sociological Construction of Tech-
    nological Systems.*
21  Callon and Latour, 'Unscrewing the big Leviathan'; Latour, *Pas-
    teurization of France.*
22  Latour, *Pandora's Hope*, p. 214.

between the human and the non-human. Marx's view that it is the practical labour of real, historical people that gives shape to their identity seems, in an even more literal sense than he had in mind, to be true.

The word 'reflexivity' also describes this relationship between identity and technology, the way human activity re-creates the material context and the material body of human existence. Knowledge and change form a reflexive circle: human systems are self-organising.[23] The way of life changes self-understanding and self-representation, and the new self-identity changes the way of life. If, earlier, people thought of human nature changing the world around them, while this nature remained stable at the centre, it is now apparent that technological activity changes human identity itself.[24] In fact, there is nothing new in principle in this phenomenon, though the scale and speed is new. It has been thought from ancient to modern times that self-knowledge is a route to self-change. 'Knowledge of the factors that have been influencing my con-duct without my knowledge does in itself open to me new possibilities of action.'[25] This, to take the obvious example, is the whole point of psychoanalysis. Thus those who welcome technological change make a persuasive point when they picture the new technologies of self and of nature as only an acceleration of existing human practice. Modern material reflexivity is not the peculiar achievement (or, depending on viewpoint, deformation) of a postmodern condition but an elaboration of the general reflexivity in human life. Mazlish, one concerned observer, has therefore attempted to seize hold of the full dimensions of this circle and, in response, to build up commitment to the human sciences in a self-conscious project to direct the future course of human evolution.[26]

The crystallographer J. D. Bernal wrote in 1929: 'We hold the future still timidly, but perceive it for the first time, as a function of our own action. Having seen it, are we to turn

---

23 See Krohn, Küppers and Nowotny (eds), *Selforganisation*.
24 For examples: on computing, Turkle, *The Second Self*; and on a consciousness-altering drug, Fraser, 'Nature of Prozac'.
25 Hampshire, *Thought and Action*, p. 190.
26 Mazlish, *The Uncertain Sciences*.

away from something that offends the very nature of our earliest desires, or is the recognition of our new powers sufficient to change those desires into the service of the future which they will have to bring about?'[27] Half-a-century later, those 'new powers' appeared to have dethroned religious and humanist notions of the self, understood as the core or being of a person, and replaced it with promises of choice between multiple selves. Bernal had recognised such a possibility and imagined a human form made up of brains detached from bodies and linked to other brains. And, indeed, some decades after he wrote, the computer transformed the interface of person and machine: 'In our contemporary writings, the relationship between machine and organism has transcended the level of mere metaphor to become one of identity.'[28] In as far as this is so, material reflexivity has collapsed the distinction between person and machine.

The sociologist of science Andy Pickering wrote: 'Machines . . . are performative agents in a sense precisely analogous to disciplined human agents . . . [I]nstead of searching for hidden causes in the domain of human motivation, say, we need to think about the explicit performances of material objects and human beings; on the other hand, we need to see human motivations themselves as bound up in all sorts of ways with machines and disciplines – as themselves at stake in practice.' Such an approach to culture specifically rejects the grand narrative of the creative human mind coming to know itself through reflection, which has been at the centre of traditional writing on the history of philosophy and science. History, rather than being 'adventures of ideas' (in Whitehead's phrase), '*is* the open-ended extension of culture in fields of agency', and the historian's business is to describe 'the specifics of this process, at particular places and particular times'.[29] Similarly, Foucault depicted history as the reshaping of patterns of power, language and practice, making possible an object of knowledge

27 Bernal, *The World, the Flesh and the Devil*, p. 96.
28 Hayward, '"Our friends electric"', p. 291.
29 Pickering, 'Cyborg history and the World War II regime', pp. 3, 4. Contrast Whitehead, *Adventures of Ideas*.

such as the human subject; he did not picture the activity of a human subject. Thus the critique of philosophical anthropology and the rethinking of the relationship of technology and knowledge have come together to question the notion of an essential human nature as opposed to a reflexive circle of knowledge and being.

Recognition of material reflexivity is embedded, even when not explicitly elaborated, in the notion of culture. The word 'culture' is hard to define in part for the very good reason that what it denotes has a reflexive character; thus the word refers to both a social realisation ('the culture') and what someone becomes ('cultured' or 'enculturated'). As a result, to describe culture in one way rather than another is itself cultural activity that fosters one culture rather than another. The Russian cultural critic Mikhail Epstein therefore concluded that 'culture is everything humanly created that simultaneously creates a human being'.[30]

Western culture at the beginning of the twenty-first century has in many significant ways embraced reflexivity, in what one social theorist called 'reflexivity in the streets'.[31] This is the reflexivity expressed in self-conscious adoption of identity through change of dress, consumption, life-style and even body. In material culture there appears to have been a shift from the search for authentic self to the re-creation of self. This brings to mind a striking passage in the young Marx's 'Economic and philosophical manuscripts' where he discussed money. The passage is condensed. But Marx laid out for inspection the idea that money is indeed the means for the realisation of being human. Each aspect of what it is to be a human has its proper material satisfaction – food, sensuous gratification, aesthetic expression, learning, even the peace of contemplative understanding. Money, in principle, makes any or all of this accessible. Marx appears to state, turning on its head the usual romantic and idealist view, that money is the means to fulfil human potential (or essence), since, with money, everything is obtainable. Postmodern culture and technology would seem to bear him out. Marx wrote:

30 Epstein, *After the Future*, p. 286.
31 Sandywell, *Reflexivity and the Crisis of Western Reason*, p. 114.

If man's *feelings*, passions, etc., are not merely anthropological characteristics in the narrower sense, but are truly *ontological* affirmations of his essence (nature), and if they only really affirm themselves in so far as their *object* exits sensuously for them, then ... *Money*, inasmuch as it possesses the *property* of being able to buy everything and appropriate all objects, is the *object* most worth possessing ... Money is the *pimp* between need and object, between life and man's means of life ... Therefore what I *am* and what I *can do* is by no means determined by my individuality. I *am* ugly, but I can buy the *most beautiful* woman. Which means to say that I am not *ugly*, for the effect of *ugliness*, its repelling power, is destroyed by money.[32]

Marx profoundly understood that this human creation, money, has become the means to human re-creation, and through re-creation it expresses what appears essential to being human. It is possible to become handsome or to have the effect of being handsome – and what is the distinction between the two if they obtain the same end? – given money. At the same time, his voice was ironic, and he was enraged by the actual inhuman consequences of money, consolidated as private property and capital, and lack of money for those who must sell their labour. In this ambivalent discussion of money – it gives and it takes away what is human – Marx struggled with the material implications of reflexivity. If a beautiful woman can be bought, no man needs to be really handsome. In such circumstances, which we face at least as acutely as Marx, where does the authority come from to differentiate what matters, or has significance, from what does not?

### Reflexivity in the human sciences

Psychologists and social scientists characteristically observe people or groups of people in the way natural scientists observe the physical world. There is no difference *in method* between the natural and social sciences. But there appears to be an obvious difference between a scientist's relationship with a stone or plant and with a conscious person or a culture: the person or culture changes in the encounter. To quote

32 Marx, 'Economic and philosophical manuscripts', pp. 375, 377.

Epstein again: 'The cultural sciences may be distinguished from the natural sciences in that the former play a key role in constituting their subject matter.'[33] This section examines how writers have understood this difference. Later, I shall argue that it is not possible, in the last analysis, to sustain belief in a difference between the natural and the human sciences on these grounds.

Humanistic philosophical writing commonly notes that the manner in which people understand themselves has consequences for what they do, indeed, for what they 'make of' themselves. Berlin wrote: 'Men's beliefs in the sphere of conduct are part of their conception of themselves and others as human beings; and this conception in its turn, whether conscious or not, is intrinsic to their picture of the world.'[34] As Iris Murdoch expressed it, 'man is a creature who makes pictures of himself and then comes to resemble the picture'.[35] In Heidegger's terms: human Being 'has a relationship towards that Being – a relationship which itself is one of Being'.[36]

These statements invoke a contrast between the human domain and the natural world, between a sphere in which knowledge re-creates the knowing subject and a sphere in which it does not. Stones do not study stones, or even animals animals, but humans do study humans. The division has indeed attracted a number of scholars. 'Reflexivity is the favorite candidate for the property which makes the human sciences unique. Reflexivity is the property of the objects of a scientific inquiry also being the subjects who carry out the inquiry.'[37] It then appears a small and natural step to anticipate that once scientists study humans like other entities in the world, as

33 Epstein, *After the Future*, p. 287. For a formal statement: Giddens, *New Rules of Sociological Method*, pp. 185–6.
34 Berlin, 'Does political theory still exist?', p. 13; see Bernstein, *Restructuring of Social and Political Theory*, p. 61.
35 Murdoch, 'Metaphysics and ethics', p. 75.
36 Heidegger, *Being and Time*, p. 32; see also Lawson, *Reflexivity*, pp. 71–3.
37 Flanagan, 'Psychology, progress, and the problem of reflexivity', p. 375.

objects which we can observe like any other objects, then the problems of reflexivity will be found not to have been problems at all. The difficulties caused by consciousness trying to reflect on itself will evaporate in the hard light of scientific realism. Opposed to this line of argument, however, are the scholars who say that the social sciences and the humanities cannot become natural sciences precisely for the reason that human activity involves reflection and that a new kind of reflection, new knowledge, changes the subject. Human beings 'are preformed by social forces that form their competence to understand themselves, and . . . in exercizing that ability, they alter themselves and others in the process, and alter the preformative conditions under which others, coming later, master a comparable but specifically different such competence'.[38] From this point of view, in the human sciences at least, it simply is not possible at root to separate subject and object. The very act of acquiring knowledge, even in the most rigorously controlled situation, is a change in the life of both subject and object.

The latter conclusion appears at odds with the belief that many psychological and social scientists have about what makes their knowledge objective. But the discussion here is not about whether any particular method of data collection and analysis is or is not objective, but whether it is a condition of knowledge in the human sphere that knowledge changes the subject. At times, it is true, scientists have thought it the very condition of achieving objective knowledge rigorously to separate the knowing observer and the human object so that the former does not affect the latter. In the second half of the nineteenth century, the new psychologists, such as Wilhelm Wundt, deplored the confusion of subject and object, and they thought that they had something new to offer precisely because experimental methods made it possible to collect reports on mental states which, they held, did not depend on

---

38  Margolis, *The Flux of History*, p. 165. Margolis attempted to think through a historicised epistemology of the human sciences. Also, Margolis, *Science without Unity*; Hallberg, Molander and Olausson, *Reflections of Humanities Studies*.

introspection.[39] Critics, however, argued that no reports of mental states could be objective, and they therefore presumed that psychology, if it is to be a science, must study observable features of the physical world just like the natural sciences. In the United States, behaviourists, like the polemical John B. Watson, therefore promoted psychology as the science of stimulus-response relationships. Through the middle years of the twentieth century, social scientists as well as psychologists made objective methodology *de rigueur* in their fields, and at times this amounted to a refusal to make any claim about the nature of objects or subjects but simply to make observation statements. There was, in this connection, much debate about the viability of the method called participant observation. Later, as the following chapter discusses, the weight of opinion went against both behaviourist psychology and positivist epistemology, and questions about self-knowledge and self-reflection returned to occupy a central position in the human sciences. Whatever the would-be objectivity of a human science, it was concluded, it has to take account of the fact that it is the activity of groups of people, in their interaction with what they call 'the world', which creates the kind of identity, and objectivity, which the groups actually have.[40]

Though by no means all psychologists and social scientists show awareness of the reflexive features of their sciences, there has for some time been a philosophical literature arguing that knowledge about people cannot be neutral, and cannot be predictive, since when people become conscious of such knowledge they change. People with knowledge are not the ignorant people they were before. Stuart Hampshire wrote (in 1959): 'As the knowledge that we may have of our own mental powers is reflexive knowledge, the object of knowledge and the knowing subject change and extend their range together.'[41] Charles Taylor argued succinctly: 'A fully competent human agent not only has some understanding (which may be also

---

39 Danziger, *Constructing the Subject*, pp. 17–52; Kusch, *Psychological Knowledge*.
40 Nelson, Megill and McCloskey, 'Rhetoric of inquiry'.
41 Hampshire, *Thought and Action*, p. 255.

more or less *mis*understanding) of himself, but is partly constituted by this understanding.'[42] He thought that this was an
insight pervasive in hermeneutics since Heidegger. And he
went on to argue that the human sciences cannot be predictive, because 'man is a self-defining animal. With changes
in his self-definition go changes in what man is, such that he
has to be understood in different terms.'[43] In other words, how
we live and what we say about how we live are not independent of each other. For Gusdorf, it was precisely the reflexive
circle between thought and being which gave purpose to his
whole enterprise of writing a history of the human sciences.[44]
Foucault too observed that, in the human sciences, thought is
'both knowledge and a modification of what it knows, reflection and a transformation of the mode of being of that on
which it reflects. Whatever it touches it immediately causes to
move.'[45]

The world of psychological suggestion offers a rich illustration. The nineteenth-century fascination with mesmerism and
hypnotism was in part about the reshaping of people's, and
especially women's, identity and actions by reshaping their
conscious or unconscious ideas. Belief that suggestion, 'ideas',
could indeed reshape a person, either temporarily during a
hypnotic trance, or more permanently through memory, seized
the imagination of those who took part and of observers alike.
Out of the possibility of this kind of reshaping came psychoanalysis and much of psychotherapy, reflexive practices *par
excellence*.

Partly because of the openly reflexive structure of psychoanalysis, Foucault was prepared to grant it some value as a
science, in a way he was not prepared to do for psychology.
Reflexivity, he suggested, makes possible a critical, hence
scientific, character.[46] As discussed earlier, Foucault's notion

---

42  Taylor, 'Introduction' to *Philosophy and the Human Sciences*, p. 3.
43  Taylor, 'Interpretation and the sciences of man', p. 55. See
     MacIntyre, *After Virtue*, p. 34.
44  Gusdorf, *De l'histoire des sciences*, pp. 233–4.
45  Foucault, *Order of Things*, p. 327.
46  *Ibid.*, pp. 373–6.

of the human sciences depended on the view that only in the modern age is there a reflexive subject, 'man'. In his view (in 1966), modern knowledge about being human presupposes an entity that is double, both subject and object of knowledge. 'Man . . . is a strange empirico-transcendental doublet, since he is a being such that knowledge will be attained in him of what renders all knowledge possible.' When Descartes put forward the *cogito* as pure thinking, Foucault argued, he imagined a capacity rationally and transparently to represent the world as thought. Descartes's 'I', therefore, had no density as subject and object of thought. In the nineteenth century, this changed, and the 'I' became the subject and object of the human sciences. With the new human sciences came the paradoxes of reflexivity.

> If man is indeed, in the world, the locus of an empirico-transcendental doublet, if he is that paradoxical figure in which the empirical contents of knowledge necessarily release, of themselves, the conditions that have made them possible, then man cannot posit himself in the immediate and sovereign transparency of a *cogito*; nor, on the other hand, can he inhabit the objective inertia of something that, by rights, does not and never can lead to self-consciousness.[47]

Modern knowing subjects are in a double bind: people cannot assert truth, as Descartes did earlier, as the outcome of a pure knowing *cogito*; but neither can people know themselves as if they were simply part of nature without consciousness. We have no transcendent truth, and yet we cannot take refuge from this by believing ourselves to be the unreflective being of nature.

Since Kant, the western philosopher and artist has, in Louis A. Sass's words, 'so to speak, looked out from consciousness and . . . [seen] the world glowing with its dependence on him'.[48] As a result, the modern 'I', for which the romantic artist is the heroic model, envisaged itself as the constitutor of the world; at the same time, however, reflection on this 'I' rendered it an object of study in the world, like other objects. The 'I' simultaneously, and paradoxically, faced both its freedom and

47 *Ibid.*, pp. 318, 322.
48 Sass, *Madness and Modernism*, p. 328.

its finitude. This is the situation of the peculiarly modern combination of the self-constituting subject which, through reflexive action, has objectified itself as object in the human sciences. This is represented in the double meaning of the word 'subject', which denotes both a discipline, or a discipline's knowledge and what that knowledge refers to, and a state of existence. 'The theorist as well as the practitioner of the human sciences is "in" and "of" what he investigates.'[49]

Graham Richards has forcefully put the point in the case of psychology: 'psychology' denotes a discipline, occupational area or body of knowledge, and a state of being. He therefore promoted 'Psychology' (big P) to denote the former and 'psychology' (little p) to denote the latter.[50] We may refer, for example, to Psychological knowledge of the emotions and to emotions as psychological states. I would also say that we refer, for example, to history or economics as disciplines but also to our place in history or to our economic state. Thus the term 'history' 'may mean the "historical actuality" as well as the possible science of it'.[51] 'Politics' denotes a discipline or, rather, the disciplines of political theory and of political science, and it also denotes political life. These paired denotations exist because institutionalised disciplines have come into existence as the collective form of reflection on what people actually do.

Since it is people with a psychology who produce Psychological theory, psychological states and Psychological knowledge influence each other. Some psychologists, like Jung, tried fully to accept the implications of this. Jung's classification of psychological types was part of an elaborate attempt to identify the personal, psychological element in objective science. In his view, the psychologist's personal type will affect what he or she thinks general Psychology is.[52] As Richards has pointed out, there are enormous implications for historical understanding if we seriously follow up on the reflexive character of the human sciences. 'The history of Psychology thereby

49 Ermarth, *Dilthey*, p. 94.
50 Richards, 'Of what is the history of psychology a history?'.
51 Heidegger, *Being and Time*, p. 430.
52 See Shamdasani, *Jung and the Making of Modern Psychology*, pp. 87–8.

becomes one aspect of the history of its own subject-matter, "psychology". The historian of Psychology is not only looking at the history of a particular discipline, but also at the history of what that discipline purports to be studying.'[53] The result is a feeling of being overwhelmed in contemplating the potential scope of the history of Psychology or of the human sciences. History, if this conclusion is taken on board, includes the history of knowledge, as in conventional histories of science, and also the history of people as psychological, social and language-using subjects. Conventional history assumes that people stay still; but according to the reflexive argument, people create knowledge, and this knowledge re-creates people, and so on. The history of the human sciences, potentially, therefore encompasses the history of human beings changing themselves.[54]

What this claim involves can be seen by looking at the history of language about being human or about human nature. Language is a record of the social constitution of 'the human'. Richards made this point with reference to psychological states. 'We can only talk about that which we have a language for talking about – and as far as the psychological is concerned we have no way of knowing what psychological phenomena are, no way of giving them meaning, except in terms of that language. If this is so, then we are bound to accept . . . that *changes in psychological language signify psychological change in their own right.*' Only with the creation of a social world with the concepts and language for the representation of human phenomena in psychological terms is it possible to call people psychological subjects. 'The very act of introducing such concepts changed the situation by providing people with new terms in which to experience themselves – and only *then* can they be properly said to refer to really occurring psychological phenomena.'[55] When this creation of a psychological

---

53 Richards, *Putting Psychology in Its Place*, p. 7.
54 I tentatively explored some practical options in Smith, 'The big picture'. For a clearly defined option, related to the interests of Psychology, see Danziger, 'Concluding comments', pp. 220–2.
55 Richards, *Putting Psychology in Its Place*, p. 9.

language occurred is a very important, complex and interesting historical question, but it must be answered elsewhere.[56] What I am stating now is that, if knowledge is reflexive, psychological states (or, indeed, social, political or economic states), as well as psychological knowledge, are historical entities, not natural or essential features of human existence. 'Our psychological classifications are constitutive of our mental states and events. Our psychological vocabulary does not classify mental states and events that exist wholly independently of the vocabulary.'[57]

This kind of argument is at the heart of a number of critical reflections on the psychological and social sciences by scientists in those fields. Here we come to another use of the word 'reflexivity', to refer to the way some scientists hope to bring their own disciplines to reflect critically on both the intellectual and political conditions in which they operate.[58] Encouragement to reflect is encouragement to change, since the reflective act in some small measure reconstitutes the field. And, in the attempt to find ways to open up reflection, studies in historical psychology or historical sociology, studies of the reflexive relation of the changing subject matter of knowledge and changing knowledge, have proved of particular value.[59] Within the social sciences, there is something of a minor industry in analysing the role of social scientists themselves in the social processes which they study. (I illustrate this in a moment in connection with the concept of society.)

Here we come face to face with the basic reason why the human sciences and the history of the human sciences are so ill-defined: their subject matter and the knowing subject are parts of a reflexive process, and as a consequence the fields

56 For the historiography, see Smith, 'History of psychological categories'.
57 Kusch, *Psychological Knowledge*, p. 248.
58 For example, Gouldner, *Coming Crisis of Western Sociology*; Morawski, 'Self-regard and other-regard'; Morawski, 'Reflexivity and the psychologist'.
59 Staeuble, '"Psychological man"'; Gergen and Gergen (eds), *Historical Social Psychology*.

are, *in principle*, not bounded. In practice, social conventions place boundaries around the subject matter of different disciplines, as they must, as the price of achieving agreement, detail and precision. Nevertheless, these boundaries are especially vulnerable to passage in the human sciences. Indeed, in the social sciences many practitioners would be hard put to say, and some would not want to say, where the boundaries of their field lie. The question of how boundaries actually do form is an important topic for the historian, and I will have much to say about it in the next two chapters.

Before going down that road, however, I want to illustrate the argument that basic categories in human self-description change and human life changes along with them. The philosopher Peter Winch stated that 'a new way of talking sufficiently important to rank as a new idea implies a new set of social relationships'.[60] As a result, can it be said that such things as mind, society, economy, culture, science, religion, polity and law existed before a language for them existed? It must be admitted that there is great scope for disagreement about how radically to pose an answer. The claim that there was no such thing as society before the nineteenth century, but only particular forms of collective existence like civil society, might not occasion too much opposition. However, the claim that there was no mind before, say, the late seventeenth century, will.

## Concepts of gender, society and race

The word 'gender', appearing alongside or in place of the word 'sex', spread with the modern wave of feminism, beginning in the late 1960s. There clearly is a modern need to separate a classification of people according to acquired social identity and assigned social role from a classification of people according to biological identity and reproductive capacity. It was essential to feminist arguments to demonstrate that social role or social position is not a direct consequence of biology. The introduction of a new category, gender, was therefore

60 Winch, *Idea of a Social Science*, pp. 122–3.

explicitly part of a reflexive process in which, as feminists hoped, thought and human identity would change together.

Beyond this, there was everything to argue for. It became clear in the 1970s that women's views markedly diverged, between those who wanted to re-evaluate 'natural' female identity and to assert that it is not only different but in many respects superior to male identity, and those who stressed the construction of all identities, female and male alike. The former position at times supported the idea of an essential womanliness, but it did so in order to elevate, not depress, women's social position. Most radically, there were arguments that essential differences of gender exist in the practice of science, in the fabric of language and in the structure of rational thought. Feminist critics of this orientation, however, suggested that any emphasis on natural difference offers hostages to fortune, and they instead pointed to the social and historical construction of all science, language and reasoning – while certainly sustaining analysis of how often these have been thought to be male preserves.[61] There were thus alternative proposals about the way to secure a positive evaluation of women's identity: through identification with a natural, or perhaps ancient, potential in women, or through the construction of a new woman. In either case, and this is the point here, descriptions of identity were enmeshed with the everyday politics of human relations. Feminists rightly understood that language makes a difference, that there is no neutral classification of people, and that the two change together. Women have, to repeat the phrase, lived a 'reflexivity in the streets'.

One interesting suggestion was that questioning the possibility of absolute foundations for reason, in the way accounts of reflexivity question it, is itself a kind of feminisation of philosophy. Such is the association of reason with profound and rigorous men, like Plato, Descartes and Kant, that the quest to found human thought on rational, timeless principles, which these men attempted to do, appears masculine. By contrast, sympathy with unfounded presuppositions at the basis of thought and with indeterminate meanings – with 'flow'

---

61 See Anthony and Witt (eds), *A Mind of One's Own*.

– appears feminine. The search for a unified order appears male, living with diversity appears female.[62] Such comments do not assert that thought is essentially gendered, but they do indicate the extent to which gender colours reflection on what the possibilities of thought are.

Historical work was extremely active and strong in these debates. In the first place, empirical research uncovered a huge amount of evidence about women's lives and achievements; it made women not just visible but active in historical narrative. Secondly, historians examined the multitude of ways, symbolic as well as literal, in which belief has structured social worlds with female subordinate to male. There is a substantial literature on the history of knowledge of sex (and sexuality – the concept originated around 1900) and its representation as a determining force in people's lives. There are also studies of how assumptions about the nature of sex have shaped knowledge; well-researched instances include primatology and psychological testing in the United States.[63] Lastly, historical work has been important to awareness of how belief about sex and gender works within occupations, notably the academic profession and the sciences. 'To inquire into what social relations are inscribing us or to name what gender politics sustain our professional standards is to engage in self-conscious reflexive thought.'[64] As a result of all these kinds of work, there is, at least in some academic and artistic circles, a high degree of self-consciousness about the way descriptions of difference have operated in different social worlds.

The very notion of society itself also has a history. Writers in the social sciences have investigated this as part of the larger project to understand the place these sciences have in modernity. It is commonly agreed that the social sciences came into existence when 'a conceptual and epistemic revolution took place which was coterminous with the formation of

62 Lloyd, 'Maleness, metaphor, and the "crisis" of reason'.
63 Haraway, *Primate Visions*; Morawski, 'Measurement of masculinity and femininity'.
64 Morawski, *Practicing Feminisms*, p. 117. See Carroy, Edelman, Ohayon and Richard (eds), *Les Femmes dans les sciences de l'homme*.

the political and technological practices that we have come to
associate with the world of modernity'.[65] Historians of sociology like David Carrithers and Johan Heilbron have argued
that sociology as a disciplinary area originated with the
conceptualisation and empirical delineation of society as a
distinct object, requiring study in terms proper to itself. This
happened in the period between Montesquieu's *De l'esprit des
lois* (1748) and Durkheim's *Les Règles de la méthode sociologique*
(1895), the period when modern forms of social organisation
also took shape.[66] At the beginning of the eighteenth century,
the French word '*société*' denoted either the fashionable *monde*
or a voluntary association of people. English-language writers
referred to 'civil society' and to 'society' meaning polite society. There was, however, no general, abstract conception of
society, subsuming political, ethical, economic and other
dimensions. Gradually, 'political and moral phenomena were
redefined as "social" phenomena, which led to a more general
and more abstract mode of thinking'. This involved a change
in occupational structures and not only ideas. 'At the same
time new groups laid claim to their own competence. They
were no longer theologians and jurists, but writers and *philosophes* who professed an accurate interpretation of human
communities.' A leading case in point is Rousseau, who
attempted a radical reshaping of notions of community and
state. In the nineteenth century, Comte became 'the first to
state the necessity for a relatively autonomous social science'.[67]
By the end of that century, an entity, society, denoting something to which people by definition, rather than out of choice
or chance, belong, was at the centre of a literature which
more and more authors identified as sociology. Durkheim then
systematically spelt out the requirements for the proper
explanation of society as a distinct object and entrenched
them in an institutionalised programme of research. In this

---

65 Wittrock, Heilbron and Magnusson, 'Rise of the social sciences',
   p. 2.
66 Carrithers, 'The Enlightenment science of society', pp. 234–6;
   Heilbron, *Rise of Social Theory*.
67 *Ibid.*, pp. 93, 248.

programme there are specifically social facts, and it is these facts – not those, say, of biology, psychology or politics – which are necessary to a science of society.

The key transition in the creation of the discipline of sociology was the shaping of society as a distinct object of cognitive enquiry, an object requiring understanding in its own terms and not as a biological, legal, political or theological entity. It is therefore possible to conclude that 'the Enlightenment invented society as the symbolic representation of collective human existence and instituted it as the essential domain of human practice'.[68] In particular, the differentiation of social thought depended on identifying and naming different social groups, institutions and interests independent of the state, thus distinguishing social as opposed to state action. The emergence of a clear idea of individual interests, of individual people with concrete motives, joining together to be active in political and economic spheres, appears to have been decisive. Where there was discussion of the individual acting in her or his own interest, there was also recognition of social groupings, institutions and *social* activity independent of the state or of religion. With the individual increasingly the focus of attention in the ordering of collective life – with individualism – the category of society developed in the middle ground between the individual and the state, and it was conceived as the place where individual and collective interests might be reconciled.[69] Thus the notions of the individual and of society developed together. Social thought expressed, and was a set of tools for, the administration and governance of the New World, the modern age, in which individuals act. It is therefore misguided to separate the history of social philosophy and social action, social science and social policy, and to regard action as the application of social thought by particular interest groups. Rather, the action and the science developed together.

Rousseau, Comte and Durkheim all wanted to reinforce social bonds. Their notions of society as a possible object of knowledge were normative, formulated in the light of what

---

68 Baker, 'Enlightenment and the institution of society', p. 96.
69 *Ibid.*, pp. 110–20.

they hoped society could become. In the background was an eighteenth-century genre which did not differentiate literary, psychological and social ideas, and in which authors held up models and ideals of life for the reflexive attention of readers, readers who would shape their own lives in the light of what they read. A distinct genre of social science began to take shape only with the development of expertise in collecting, statistically analysing and presenting data as the means to create regularity and order in social life. Even so, when authors of social science responded to the new conditions of modern life with new claims to expertise, their projects remained moral ones.

In sum: 'Social science, as we know it, developed in modernity, as part of . . . societies' self-reflective knowledge.'[70] Part of this self-reflective knowledge became social science's knowledge of itself, when, after about 1970, reflexive knowledge became widespread in social theory. At this time, Alvin Gouldner, in reaction against the then dominant positivism in social science in the United States, called on his fellow sociologists to become systematically reflective about their own role as social actors.[71] Subsequently, almost everyone came to agree with Anthony Giddens's observation that 'social science is actively bound up with its "subject matter", which in some part it helps reflexively to constitute'.[72] As a result, Giddens argued, 'reflexivity has to be reconstructed within the discourse of social theory not just in respect of the members of society whose conduct is the object of study, but also *in respect of social science itself as a form of human endeavour*'.[73]

That the word 'race' is reflexively related to the object which it is held to denote, that is, word and object came into existence together and faded away together, is particularly clear. Or, perhaps I should say, modern liberally minded people who

---

70 Wagner, *History and Theory of the Social Sciences*, p. 81.
71 Gouldner, *Coming Crisis of Western Sociology*.
72 Giddens, 'Preface', in Wagner, Wittrock and Whitley (eds), *Discourses on Society*, p. xiv. Also Sandywell, *Reflexivity and the Crisis of Western Reason*, pp. 1–51.
73 Giddens, *Central Problems in Social Theory*, p. 47.

have become very self-conscious about words think it clear. As with 'sex', such people have taken to heart the aphorism that to learn 'the distinction between what is necessary and what is the product merely of our own contingent arrangements, is to learn the key to self-awareness itself'.[74] It is a principle of modern rational individualism: 'Rationality, and a choice between alternatives that is genuinely one's own choice, presupposes a full knowledge of the hitherto unrecognised causes of the confinement of one's choice to a particular range of possibilities.'[75] In other words, if we are to be free to chose the kind of world we want, we need to understand the particular ways we classify what is in the world and how we describe it in one way rather than another.

Descriptions of race have, more visibly than most descriptions of people, had consequences. Here, in the history of the idea of race, there is impressive evidence how belief and life together reflexively create human – and, disastrously, inhuman – subjects. Why should anyone classify people in terms of race? I take the view, which I think widely shared, though some biologists and psychologists disagree, that the word 'race' does no work as a description which could not be done better by other words and other ways of describing the differences between human groups and populations.[76] Few people who use the word appear able to say precisely what they mean by it; yet the word is used pervasively.

Earlier English-language writers most often used the word in the expression 'the human race'. This usage continued alongside reference to different 'human races' through the nineteenth century and later. Prior to this, however, in Renaissance and early modern Europe, authors who sought to describe difference turned to the Ancient Greek distinction between civilised and barbarian peoples. Observers from Montaigne to Locke commented that what, in European eyes, distinguishes people in the New World is what is lacking in their languages,

---

74 Skinner, 'Meaning and understanding in the history of ideas', p. 67.
75 Hampshire, *Thought and Action*, p. 256.
76 For a modern assessment, Malik, *Meaning of Race*.

customs and goods, not their race. In the eighteenth century, writers recreated the contrast between savage and civilised as a tool of critique, using the contrast between peoples, like the contrast between different times and places, as a vantage point from which to comment on the artifice in contemporary mores and political affairs. At the same time, descriptive history branched out into the comparative study of languages, jurisprudence, customs, antiquity and politics in all their diversity, as well as into the natural history of plants, animals and people. A biological conception of different races then started to develop late in the century, especially when comparative anatomists began to claim that they had rigorous observational methods to distinguish human types on the basis of bodily characteristics. Throughout the nineteenth century, anatomists claimed scientific superiority, that is, greater precision and objectivity, for their criteria of difference over the criteria worked out in comparative philology and ethnology. When they classified differences, however, they re-created as scientific knowledge pre-existing assumptions about the hierarchy of higher and lower qualities of people. The biological approach to difference also brought argument over whether human types belong to one or to several species to the centre of attention. And, after 1850, in the period of European nationalist struggles, of vigorous expansion of the colonial system and of slavery, civil war and reconstruction in the United States, the biological theory of race achieved tremendous prominence. In addition, evolutionary theorists re-expressed the hierarchy of difference represented in classifications of race as a historical narrative of progress.

The prominence of the word 'race' in scientific and common language alike makes it all the more striking that at no time was there agreement about definition or the actual number of races. There were also always scientists, like the Berlin physician and politician Rudolf Virchow, critical of the descriptive basis of race claims. The twentieth-century discovery of blood groups and development of molecular biology, along with a reassessment of population thinking in evolutionary theory, suggested much more rigorous ways to classify people. During the 1930s in the United States, there was a very revealing shift, notably evident in psychology, from the study of race to the study of racial prejudice.[77] After 1945, with the consequences of race-thinking all too apparent everywhere, in the Holocaust, and from the Japanese imperium to the segregated US army, a majority of social and natural scientists found it at one and the same time possible and desirable to do without the category of race.

The mainstream scientific response to racial thinking in science has been to dismiss it as bad science or as prejudice masquerading as science. The criticism, as found in Stephen Jay Gould's standard account, took two forms, both of which, I note for my purposes, exemplify the empirical rather than reflexive language characteristic of natural science.[78] First, claims about race were value-laden and not objective; and second, the methods used to determine racial difference were flawed. Recent natural scientists have therefore concluded that racial science was not science at all. This argument is certainly well founded. It is a different point, however, to use the concept of race, as I am doing, like Richards, as a particularly vivid example of the reflexive relationship of belief and practice.[79] This point leads to a recognition, which the empirical arguments do not, that to indict race ideas for being bad science expresses an antagonism to racial thinking, just as earlier racial science expressed support for ways of life giving race priority. The opposition to race descriptions, like support for them, exists as part of a reflexive relation between knowledge and its objects.

Because descriptive terms about human subjects are at the same time forms of social action, no language is neutral. The history of the idea of race is a history of ways of life in which it became possible to create races as part of the world; and this history must now include the history of how it became thinkable to exclude race from the world. Any human self-description is a determination of identity with consequences for action; action, in its turn, creates new self-descriptions. Feminist writing has made this very clear. Hence the search

77 Samelson, 'From "race psychology" to "studies in prejudice"'.
78 Gould, *Mismeasurement of Man*.
79 Richards, *'Race', Racism and Psychology*.

in the philosophy of the social sciences for adequate ways to conceptualise belief and action as elements of a common process. As Dewey argued: 'Knowing is itself a mode of practical action and is *the* way of interaction by which other natural interactions become subject to direction.'[80]

As these sketches of the notions of gender, society and race attest, there is a rich literature, at once historical and belonging to the human sciences, about how and why notions of group identity and difference came into existence. Scientific work elaborates categories into detailed objective descriptions. Whatever the detail and precision, however, all scientific description, whether of society, race or gender, builds theoretical presuppositions into its most basic terms. This is the import of the theory of knowledge outlined in this chapter. Which presuppositions people actually adopt is the outcome of social and historical processes, like discipline formation in the sciences, like shifting power relations between different groups of people and like changes in authority over knowledge in daily life. This argument makes historical knowledge essential to understanding how and why we think in the way we do. But, before developing this thesis, I turn to the history and philosophy of relations between the natural sciences and the human sciences.

80 Dewey, *Quest for Certainty*, p. 104.

# 3

# Relations of the natural and human sciences

## The standard model

In January 1829, in his rooms on the Left Bank in Paris, Auguste Comte began a course of lectures to a select audience. Published as the *Cours de philosophie positive* (1830–42), these lectures laid out a philosophy of knowledge of humanity, unadulterated by theology or metaphysics, as the means to redemption. Comte subsequently founded a positivist church, modelled on Catholic institutions, in order to inspire others with his vision. Though this church met with only some success, belief that people can know through science what they need for the good life took hold and, in important respects, lasted through an ever more disillusioned twentieth century. In 1889, the Brazilian Republic inscribed the positivist slogan *'Ordem e Progresso'* on its flag. John Stuart Mill, who both supported and criticised Comte, published *A System of Logic* (1843), which remains an authoritative source for the view that inductive argument from facts to general laws is the manner of reasoning across the human and natural sciences, and that on such reasoning all else depends.

Comte was the most systematic proponent in the nineteenth century of the view that there is one kind of objective knowledge of the world, and that the natural sciences have shown in practice what this is. The sciences of man (in the plural), he wrote, will create positive knowledge by 'tracing the course actually followed by the human mind in action, through the examination of the methods really employed to obtain the exact knowledge that it has already acquired' in the physical sciences. What differentiates positive and, in his view, certain knowledge from the false claims of theology and metaphysics, is that it is restricted to knowledge of observed regularities. 'It is the nature of positive philosophy to regard all natural phenomena as subject to invariable natural *laws*, the discovery of which . . . is the aim and end of all our efforts . . . we do not pretend to expound the generative *causes* of phenomena.' We do not need, he argued, to speculate about either transcendental forces or unseen natural causes in order to know what we need to know in order to bring order to human affairs. We need only to know the laws which events follow, in the human sphere as in physical nature, in order to act wisely. There will be social harmony once people accept these positivist principles: 'if the union of minds in a communion of principles can once be established, suitable institutions will necessarily spring from it'.[1] Later positivist philosophers, notably the members of the Vienna Circle (who initiated logical positivism in the 1920s), did not follow Comte's religion, but they did share the essentials of both his theory of knowledge and his hopes for the rational organisation of human life. Moreover, even after strict positivism was thought untenable as a theory of knowledge, in the years after 1950, many natural and social scientists perpetuated an image of the physical sciences as an ideal to follow and of science generally as the motor of human progress.

An empiricist, if not strictly positivist, theory of knowledge became characteristic, and at times compulsory, in the twentieth-century human sciences. For the founders of the modern social sciences early in the twentieth century, it was *methods* to make empirical knowledge rigorous and systematic which were needed to ensure the position of the new sciences in competition with ignorance and prejudice. Sound methods, they thought, would also allay the suspicions of physical scientists that social science was not real science. Already in the nineteenth century, such beliefs had had influence in breaking down boundaries between the study of physical nature and the study of human beings. For many nineteenth-century

---

1 *The Essential Comte*, pp. 32, 24, 38. See Braunstein, 'Philosophie des sciences d'Auguste Comte'.

researchers and their descendants in the emerging disciplines of psychology, anthropology, sociology, economics, linguistics and political science, it was self-evident that progress depends on following the example of the physical sciences. Like the Victorian mathematician and philosopher W. K. Clifford, they envisaged the domain of the natural sciences expanding to become the domain filled by 'all possible human knowledge which can rightly be used to guide human conduct'.[2] This stated a principle of naturalism, which a modern philosopher of the social sciences succinctly described as the 'conviction that the aim of the social sciences is the same as that of the natural sciences'.[3] The term signals 'the refusal to take "nature" or "the natural" as a term of distinction', separating the realm of being human from the realm of nature.[4] Evolutionary theorists in general, and Darwin in particular, persuasively established empirical demonstrations of this continuity. What I call the standard model of knowledge in the human sciences therefore appears to be the straightforward extension of natural science to encompass human life as part of nature.

Many psychologists and social scientists became convinced that only lack of skill, persistence and rigour, and of course money, held up delivery of reliable observations of the regularities in the, admittedly complex, realm of human conduct and affairs. It may have been extreme, but it is symbolic, that the psychologist Clark L. Hull, who was head of Yale University's Institute of Human Relations in the late 1930s, kept an open copy of Newton's *Principia Mathematica* on display in his office. Visitors were to be quite clear about the model for the project under way. Numerous social scientists of Hull's generation admitted that they had secured little by way of authoritative knowledge and that they could not be certain of the effects that social science informed policy would have. But they confidently believed that this reflected only the late start and the relative immaturity of their science. They had faith that the method for securing knowledge was available

2 Clifford, 'Body and mind', p. 736.
3 Bernstein, *Restructuring of Social and Political Theory*, p. 43.
4 Randall, 'Epilogue: the nature of naturalism', p. 357.

and that, with time and resources, the human sciences would take their place alongside the physical sciences and, like the latter, change the world. In the United States, during World War II and after, politicians agreed to a massive expansion of the psychological and social sciences, and other western countries followed suit. The image of science as materially successful in the defeat of Germany and Japan and morally successful in the triumph of reason over barbarity supported the expansion of science in the human sphere. This expansion continued, seemingly without intense scrutiny, until the early 1970s, often to the accompaniment of rhetoric expounding the virtues of positivist methodology.

In the 1930s, immigrant and refugee logical positivists from Europe to the United States created the unity of science movement, which attempted to build a community committed to truth around a formally stated scientific method. The degree to which this philosophical movement actually influenced the course of the sciences is questionable, but, for a while, it upheld a social as well as philosophical ideal of objective science as the leading edge of progressive humanity. Something comparable was present in the Soviet Union, though in this setting spokesmen for science anathematised positivism as incompatible with Leninist realism and upheld dialectical materialist philosophy as the basis for the unity of the sciences. In practice, the identification of Marxism-Leninism with the interests of the Communist Party of the Soviet Union pre-empted the development of the social sciences as this occurred in the West; the last thing the Party could tolerate was empirical study of its own activity in society. Yet, in other areas of the sciences, the circumstances reinforced the extremely empiricist inclinations of many individual scientists. Struggling to maintain personal integrity and the status of the intelligentsia in Russia, scientists staunchly believed in the real objective standing of scientific knowledge – the knowledge of which they, not the state, could claim to be the authoritative source.

Positivist epistemology, formally understood, was in decline for philosophical reasons from the 1950s on, and criticism spread from formal epistemology to become a much broader questioning of practices within the psychological and social sciences. Critics attacked the view, stated in the words of one spokesman for the rationality of science, that 'science aims at an observer-independent account of the world, transcending human meaning, culture and ideology'.[5] The most obvious objection to this is that it is not possible to state facts independently of theoretical presuppositions.[6] The natural philosopher Whewell long ago made the key point: 'And thus, a true Theory is a Fact; a Fact is a familiar Theory. That which is a Fact under one aspect, is a Theory under another.'[7] If all observations are theory-laden, it is not possible to make sense of an ideal of science 'transcending human meaning'; what the facts are in any particular domain depends on the nature of theories and concepts in that domain. This is a conclusion with great importance for scientific research as well as for philosophy. A second, related line of argument holds that there is no single scientific method for deriving knowledge from empirical facts – a point that many scientists, like Peter Medawar, have been happy to agree with.[8] A third line of opposition to positivism in the human sciences took the view that knowledge about people does not have the same form as knowledge in the natural sciences. This argument led to a debate about reduction, about whether it is possible to demonstrate that psychological and social science explanations are at base explanations in terms of physics, chemistry and biology. A strongly pro-reduction position dismisses any claim to knowledge not ultimately formulated in terms of the causal laws of the physical world. A large literature in the philosophy of the social sciences grew up in opposition to this, which I explore in the following section. Lastly, critics of positivism had even larger game in their sights: the notion that there was, or could be, value-free knowledge. I will return to this later, as it is fundamental to argument about the value of historical knowledge in the human sciences.

5 O'Hear, '"Two cultures" revisited', p. 5.
6 See Hesse, 'Theory and observation'.
7 Whewell, *Philosophy of the Inductive Sciences*, vol. 1, p. 40.
8 Medawar, *Art of the Soluble*.

## The explanation of human action

There is a well-established critique of the standard model, that is, the explanation of human action in the same terms as the explanation of physical events. Philosophers who think the debate passé may simply move on. But, though this critique goes back a number of decades, it has become very pertinent again in the light of evolutionary and neuroscience claims to understand the whole range of human activity. What is thought well established in one area of learning (the philosophy of social science) may be ignored in other areas (biological science) and hence bear recapitulation.

Beginning in the late 1950s, and continuing through the 1960s, English-speaking philosophers and social scientists profoundly criticised the view that all explanations, of human as well as physical phenomena, have the same form. To be sure, many continental European intellectuals, and a number of Anglophone ones as well, had never been persuaded in the first place by naturalism, but their writings now began to appear mainstream rather than marginal. New authors, like Winch and Taylor, attacked behaviour as a category in terms of which to unify explanation, in mechanistic terms, of all the kinds of things people do. The target here was the hope, originating in the United States, of unifying the subject matter and methods of the psychological and social sciences, including political science, under the rubric of 'the behavioral sciences'. And, it should be noted, whatever the criticisms then made of this unifying ideal, they have not prevented the rise of new unifying ideals, as in the programme, to which many neuroscientists appear committed, to explain all human action in terms of brain functions.

The critical point of departure was the language which we use when we say that we understand why someone or a group of people does something. To ask for an explanation of a human action is to ask for a reason, understood as a purpose or intention, why the action occurred. The natural sciences do not provide such explanations but instead explain by reference to physical events understood as causes. If physiology explains the play of impulses through the nervous system when someone goes out of the room, everyday explanation ascribes the events, for example, to personal feelings and social conventions that cause the person to take offence. The anti-naturalist position is that no amount of knowledge of observed regularities in people's movements, and no amount of knowledge of the nervous system, will tell us what we want to know, which is about feelings and conventions. No amount of knowledge of physical nature, the argument goes, will inform us when we ask why a *person* acts in a certain way. Rather, we must have psychological or social knowledge of mental events and social rules. Similarly, no demonstration that an action has a certain statistical probability will answer the question why a particular person or group does a particular thing. For that explanation we need a particular history, and this history must include an account of the person's or group's self-understanding. The business of psychological and social science, the argument continues, is to give this ordinary language understanding, of what it is to explain why people do what they do, objective, rigorous and systematic form.

This critique of reduction defines human beings as agents with purposes; that is, to use the everyday expression, it treats people as having minds. Since people are embedded in history and in a social setting, they share purposes, which take the form of institutions, customs and languages. These shared phenomena are, in a word, culture. According to this argument, it is completely misguided to put forward behaviour as the common subject matter of the psychological and social sciences. Behaviour is a physical event; but what the psychological and social sciences seek to explain is culture and the way culture exists in the reasons, or intentions, of social actors. 'Explaining actions is explaining choices; and explaining choices is exhibiting why certain criteria define rational behaviour for a given society.'[9]

The development of this argument gave rise to a number of questions. Is it the case, for example, that a description of the reason why someone makes a choice actually assigns the reason as a cause of what she does? Taylor thought so: 'our ordinary accounts of action are causal in a perfectly straightforward

9 MacIntyre, 'A mistake about causality in social science', p. 61.

sense, even if they are not mechanistic'.[10] Ordinary language appears to treat reasons as causes. But, as Taylor also noticed, these human causes differ from mechanical causes in the sense that the former explain action by reference to a certain standard, value or judgement of significance (for example, when someone acts out of shame), whereas the latter do not.

Argument in the philosophy of the social sciences owed much to Wittgenstein's *Philosophical Investigations* (1953). He claimed that all human activity, most especially all language use, consists of rule-following in accordance with, or embedded in, a way of life. In his famous formulation: 'What has to be accepted, the given, is – so one could say – *forms of life*.' When we ask for an explanation, we do not ask for an explanation in general, whatever that might mean, but for an explanation according to an accepted usage of language. An explanation is an explanation in a context. As language is in its nature social, every account of the world, including a scientific explanation, is an account according to an established convention or what Wittgenstein called a 'language-game'.[11] This argument proved especially damaging to positivism. It suggested that a procedure for the verification of statements about observations by observations themselves could never decide what is to count as knowledge. A conventional rule about language use is also needed. It appeared to follow from this that the social sciences have a specific and irreducible role in understanding knowledge: they study rules. 'The different uses of language have ultimately to be understood as acts of communication, and therefore as parts of different forms of social life.'[12] This was not at all what nineteenth-century positivists had envisaged for the social sciences when they had promoted the study of social and historical facts in order to arrive at general laws.

In *The Idea of a Social Science*, one influential study drawing on Wittgenstein, Winch argued that forms of understanding in the natural and human sciences are necessarily distinct.

---

10 Taylor, 'How is mechanism conceivable?', p. 167.
11 Wittgenstein, *Philosophical Investigations*, pp. 226[e], 5[e] (§ 7).
12 Hampshire, *Thought and Action*, p. 233.

This argument, when published in 1958, strikingly contested the dominant view in Anglo-American social science. Winch argued that what we call social reality is a form of life made distinctive by the rule-following nature of a language. Knowledge about forms of life, he argued, cannot be exclusively empirical, let alone restricted to descriptions of physical regularities, since it involves analysis of how language and social action relate to each other. We cannot know the world separately from the language in terms of which we know the world. 'To assume at the outset that one can make a sharp distinction between "the world" and "the language in which we try to describe the world", to the extent of saying that the problems of philosophy do not arise at all out of the former but only out of the latter, is to beg the whole question of philosophy.' The possible implications of this for the natural as well as the human sciences, which others later took up, were not Winch's prime objective. Though he stated that 'our idea of what belongs to the realm of reality is given for us in the language that we use', his goal was restricted to the affirmation that sociology is the understanding of forms of life, and that such understanding is logically dependent on relations between people embedded in language use.[13] To establish intelligible knowledge of social life, Winch concluded, is to understand how people use language in saying what they are doing. A prominent statement for a new, anti-positivist social psychology therefore sought a science which would 'converge on the model that people have of themselves, and that is embedded in much of the logic of ordinary language'. In short: 'For scientific purposes, treat people as if they were human beings.'[14]

As Winch perceived, this view of social science knowledge made it necessary to analyse the relationship between the investigator's notion of intelligibility and form of life and the notions of intelligibility and forms of life of the subjects studied. At the very least, this obliged researchers to take seriously the way other people understand the world, however wrong

13 Winch, *Idea of a Social Science*, pp. 13, 15.
14 Harré and Secord, *Explanation of Social Behaviour*, pp. v, 84.

by the researchers' standards it might be. Other people's understanding of their world is part of the explanation for how they act and why they have the institutions that they do. With such arguments, philosophy lent authority to what was in fact common practice in many areas of scholarship about the human world. In history, biography and social anthropology, for example, as in the study of art, literature or religion, scholars, in spite of the naturalist challenge, had throughout the twentieth century commonly attempted 'to enter into' the world of the subjects they studied. Once marginal in status, judged by a criterion of positive knowledge, these disciplines began to find themselves held up as models of how to understand human action.

The work, and reception, of the Oxford philosopher and historian of the interwar years, Collingwood, is relevant here. The intense empiricist ethos of the discipline of history in Britain had made his reflections on historical knowledge, when first published, look suspiciously exotic. He argued that historical knowledge consists of the rethinking of human thought: 'The history of thought, and therefore all history, is the re-enactment of past thought in the historian's own mind.' When views about explanation in the social sciences shifted, as Winch and others had suggested they should, a new audience praised Collingwood for having seen that our understanding of other people, including people in the past, necessarily involves knowledge of how they explain their actions. This insight was at the base of the distinction he had drawn between history and the natural sciences. 'Historical knowledge, then, has for its proper object thought: not things thought about, but the act of thinking itself. This principle has served us to distinguish history from natural science on the one hand . . . and on the other from psychology as the study of immediate experience, sensation, and feeling.'[15] The new philosophers of social science were mostly unsympathetic to the sort of idealist epistemology which Collingwood had espoused. All the same, he appeared to have appreciated the significance

15 Collingwood, *Idea of History*, pp. 215, 305.

of intellectual history and the history of belief systems in human self-understanding.

Another dimension of the intellectual shift among social scientists, which also encouraged interest in disciplines in the humanities, was a new focus on *meaning*. This reanimated an older debate in German-language philosophy about the relation between the natural and social sciences and about the claim that the former explain and the latter interpret and hence arrive at understanding rather than explanation. Geertz, seeking to define the study of culture, stated: 'Believing, with Max Weber, that man is an animal suspended in webs of significance he himself has spun, I take culture to be those webs, and the analysis of it to be therefore not an experimental science in search of law but an interpretive one in search of meaning.'[16] The reference to Weber was apposite. He had defined sociology as 'a science concerning itself with the interpretive understanding of social action', and he went on to say that 'we shall speak of "action" insofar as the acting individual attaches a subjective meaning to his behavior – be it overt or covert, omission or acquiescence'.[17] In the Anglophone world of the 1960s and early 1970s, to reassert the rights of 'interpretive understanding' as opposed to explanation was a sign of reaction against the standard model in social science.

As scholars examined the explanation/understanding distinction, however, it broke down before their eyes. Since criticism of the positivist theory of knowledge centred on the impossibility of establishing theory-free observation statements, there appeared to be a sense in which any statement about the world is an interpretation; any statement presupposes one set of theoretical notions rather than another. Even an explanation in terms of physical causes relies on a prior interpretation of what is physical, what is a cause and so on. Whatever the differences otherwise between them, Anglo-American philosophers came close to Heidegger's argument that ontological statements logically precede empirical science: 'We must always bear in mind . . . that . . . ontological

16 Geertz, 'Thick description', p. 5.
17 Weber, *Economy and Society*, vol. 1, p. 4.

foundations can never be disclosed by subsequent hypotheses derived from empirical material, but that they are always "there" already, even when that empirical material simply gets *collected*.'[18]

It is not possible, at a fundamental level, to separate natural science knowledge from other scholarly knowledge because the former articulates causal explanations and the latter not, or the former general laws and the latter the understanding of particulars. Such differences are quite commonly present; but these differences are contingent and do not reveal underlying differences of rationality. As Putnam wrote, 'a better approach would be to begin by recognizing that interpretation, in a very wide sense of the term, and value are involved in our notions of rationality in every area'.[19] There is no mileage in the view that the so-called soft psychological and social sciences would somehow escape interpretation if they became like the so-called hard natural sciences. Knowledge in the hard sciences is also interpretive, since interpretation is a feature of the exercise of rationality itself.

The explanation/understanding distinction also broke down because observers found causal physical explanation in the humanities disciplines and interpretive argument in the natural sciences. For example, a mixture of multi-causal explanation and interpretive narrative is the rule rather than the exception in historical writing; and the sub-discipline of social psychology includes both experimental research on causes and the interpretive analysis of discourse.

Causal explanation involves using concepts, and it obviously belongs to the natural sciences to interpret concepts; conversely, interpreting meaning in the social world is not possible independently of causal explanation.[20] This led Paul Ricoeur, whose special interest at this point was historical knowledge, to note that he aimed 'to integrate the opposed attitudes of explanation and understanding within an overall conception of reading as the recovery of meaning'.[21] Integration is possible

18 Heidegger, *Being and Time*, p. 75.
19 Putnam, 'Beyond historicism', p. 300.
20 See Giddens, *Central Problems in Social Theory*, p. 259.
21 Ricoeur, 'What is a text?', p. 161.

because 'the nature of explanation is the same in history as in the natural sciences'. He then observed, making a point which is important for my later argument: 'The question is not whether the structure of explanation is different, but rather in what sort of discourse this explanatory structure functions.'[22] That is, what separates the natural and the human sciences may not be so much different kinds of explanation as the different purposes of explanation. Explanation serves different purposes in different discursive contexts. It is this, I suggest, which does separate the natural and the human sciences: they develop different forms of knowledge to the extent that they have different purposes.

Despite all these qualifications, the distinction between explanation and understanding had an influential place in comparing knowledge of people with knowledge of other things. One reason for this was the continuing presence of a humanist and religious sensibility which led writers to treat 'human' as a moral (or spiritual) category, not as the name for a set of empirical characteristics. According to this view, the human sciences, by definition, have a moral (or spiritual) subject matter. This appears, for example, to have been Berlin's position. For him, 'the basic categories in terms of which we perceive and order and interpret data' about people and mark how they differ from other things, are, indeed, categories; 'that is . . . they are not themselves subjects for scientific hypotheses about the data which they order'.[23] For human life to be the subject of a science, this science must presuppose the normative category 'human'. By contrast, the physical sciences presuppose a category, 'nature', which is not normative and not human. It follows that it makes sense to interpret the meaning of what people do, since by virtue of being people they act in relation to certain norms or values; and, equally, it does not makes sense to interpret the meaning of physical nature, since there can be none – the meaning of nature lies in what people say about it.

Weber's words retain their force: human culture assigns significance to the world, and the cultural sciences, unlike the

22 Ricoeur, 'The narrative function', pp. 276–7.
23 Berlin, 'Does political theory still exist?', p. 23.

natural sciences, therefore have 'significance' as their subject matter: '"Culture" is a finite segment of the meaningless infinity of the world process, a segment on which *human beings* confer meaning and significance.'[24] This basic principle, that humans alone are meaning-conferring, and that this is a categorical not an empirical statement, led the human sciences in the 1960s and 1970s to claim autonomy from the natural sciences. Numerous more recent authors, shortly to be discussed, will have none of it.

First, I return to Winch's statement that 'reality is given for us in the language that we use'. This had radical implications for understanding science, which sociologists of both natural and social science took up in the 1970s. They began to study the world of scientists as a 'form of life': we have to understand social rules and rules of language if we are to understand how scientists establish knowledge of the world. The new sociology of knowledge did not impugn objectivity as it exists in the social world of scientists. What it did question was belief that scientific knowledge is objective from some would-be absolute standpoint in the way pre-reflexive theorists of knowledge, like the positivists, had imagined possible. Some scientists responded with outrage to the suggestion that we might understand the authority that scientific knowledge has in terms of social practices rather than in terms of procedures for confirmation (or falsification) through rigorous observation. Yet the sociologists were obviously not in some wicked way against reason but for a correct understanding of the reflexive nature of reason.

This sociology of knowledge came together with other factors to place more emphasis on the *history* in the history of science. Political radicalism motivated study of the social and historical conditions mediating the language of truth claims. Historians of political theory, using the precise formulations of ordinary language philosophy, persuasively argued that only knowledge of specific historical contexts makes it possible, properly, to talk about the meaning of great texts or ideas.[25]

---

24 Weber, '"Objectivity" in social science', p. 81.
25 Skinner, 'Meaning and understanding'.

Canguilhem, very influential in France, rejected abstract philosophy of science (especially in biology and medicine) for a historical study of the concepts in fact central to established scientific knowledge.[26] The critique of positivist epistemology and empiricist theories of scientific rationality thus gave the history of science a new lease of life. If it is not possible to elucidate the entire meaning of scientific statements in empirical terms, this is because meaning is historically constituted. If observation is theory-laden, then knowledge is historical. This conclusion made historical research essential to the study of the natural and social sciences as human activities. By contrast, most scientists had hitherto regarded history as a possibly interesting but, finally, decorative adjunct to science. The collapse of the positivist programme led to a decisive upgrading of the status of historical knowledge.

Knowledge in the physical sciences, which had been held up as the model for the social sciences, turned out, on closer inspection, to require understanding in terms which originate with knowledge in the social sciences and humanities. Many natural scientists, along with many empirically oriented social scientists, turned away in incomprehension from such conclusions or rejected them outright. The conclusions appeared to offend against belief that scientific work produces empirical knowledge of what is real. The theory of knowledge which I have been sketching, therefore, in practice, tended to reinforce disciplinary divisions in universities and gave new expression to the notion of 'the two cultures'. Natural scientists who thought the world defined for us by physical and biological events, and social theorists or literary scholars who thought language defines the world, largely ignored each other.

## The new materialism

Since the 1980s, there has been a marked opening up of a fully naturalistic approach to mind, language and culture, once again encouraging belief that natural science knowledge is the base of human self-understanding. Evolutionary biology

26 See Braunstein, 'Bachelard, Canguilhem, Foucault'.

and neuroscience have been of special importance. It is not to the point now to assess these sciences, but I will respond to some of the claims made about neuroscience, because it is here perhaps that arguments to make human action fully explicable in terms of physical causes are most articulate and influential.

A number of brain scientists and philosophers of mind have declared clearly and emphatically that there is no separate mental stuff; further, and to my mind more importantly for philosophical argument, they have declared that knowledge of brain events explains what goes on when people experience or act. Many, but not all, treat 'mind' as the generic term for a number of brain functions. Sensible qualities (like colour, brightness or warmth, called qualia), feelings and the reasons that people have, or the intentions they state, are all functions or properties of brain events. (Philosophers differentiate 'function' and 'property' in arguing about these questions.) This goes against the whole tenor of the philosophy of social science outlined in the previous section.

Computational modelling has been a defining technology throughout the course of the debates. The technology has shaped the direction of thought and research, since computing suggests how events in a physical system can result in reasoning or using information to produce relevant courses of action. All the same, there has been, and remains, considerable disagreement about whether computing can in any way model conscious qualities as properties of a physical system. The most emphatic proponents of computational approaches argue a position called 'eliminative materialism' and believe that scientific knowledge will replace everyday language ('folk psychology') about why people do what they do.[27] The debates have left consciousness as an acute problem. Thus, the philosopher of mind John Searle, in opposition to the 'eliminativists', has taken the naturalistic view that consciousness is a property (not function) of brain states which intrinsically have what he called 'a first-person ontology' – these brain states just do have subjectivity. He therefore rejected materialism

27 Churchland, 'Eliminative materialsm'.

and the elimination of ordinary language descriptions of mind, though he also firmly treated mind as a subject for biological knowledge. As Searle wrote: 'If we try to draw our own consciousness, we end up drawing whatever it is that we are conscious of. If consciousness is the rock-bottom basis for getting at reality . . . we cannot get at the reality of consciousness in the way that, using consciousness, we can get at the reality of other phenomena.'[28] Such an admission leads, it would seem, to the questions raised here in connection with reflexivity (though this is not the way Searle, an emphatic realist in epistemology, saw things). Both Searle and the materialists who opposed his views believed, however, that we must build our understanding of mind and culture on knowledge of the brain. It is this that I am calling the new materialism, or, more accurately in relation to a philosopher like Searle, the new naturalism. It is a position which claims, whatever value or meaning the humanities disciplines or ordinary people find in individual and collective expressive life, that it is natural science which, at base, explains it.

It is striking that some philosophers of mind have joined forces with neuroscientists to make these arguments. Through the middle decades of the twentieth century, philosophers and scientists generally agreed to keep separate logical claims about concepts, reasoning and language and empirical claims about the world. This was a division entrenched in intellectual thought since Kant and given new impetus by the rigorous empiricism of the natural and social sciences. It was thus possible, for example, for Berlin (quoted earlier) to define being human in terms of moral agency and to treat this definition as a conceptual matter and not a matter for science. The new naturalism, however, has opposed this intellectual division of labour.[29] It is part of the naturalistic argument that the conceptual sphere is itself a product of nature and we must therefore study it as we study nature. This argument collapses the conceptual and the normative into the natural. Many philosophers strongly oppose this step, and for reasons

---

28 Searle, *Rediscovery of the Mind*, pp. 16, 96–7 (italics removed).
29 See Churchland, *Matter and Consciousness*.

that lie in the complexities of argument about how we reason at all. There are no easy resolutions of such controversies, though I certainly hope that this book is building up a case for seeing the limits of neuroscientific explanation.

Scientific materialism, to be successful, must overcome a host of difficulties, of which, as many critics believe, the absence of any even remotely plausible approach to the physical explanation of sentience, or consciousness, is the most tangible. In addition, the arguments developed earlier, critical of behaviourism and other forms of reduction to mechanistic explanation in psychology and the social sciences, still retain their force. Whatever the empirical advances in neuroscience actually are, it is still possible to argue that the understanding we seek about human subjects fundamentally, and irreducibly, concerns meaning, values, social rules and the expressive world made possible by language. As a matter of principle, no material, causal theory can replace the search for knowledge of meanings, values, rules and language without a total change in our form of life and in the significance that life has for us. It is, perhaps, possible to imagine a form of life in which nothing has meaning, there are no rules or conscious judgements of significance and there is no language. But it is not the present human world. If enough people were to take some neuroscience claims literally and exclude all other forms of knowledge, it might perhaps *become* the human world. One of the consequences in this imagined world, however, would be that the pursuit of neuroscience knowledge, or of scientific truth, would become as meaningless as anything else. But, if we reiterate the ordinary language argument against this kind of materialist explanation and materialist future, we keep alive a way of life in which we understand people by reference to judgements and intentions. The materialist philosophy of mind, like any other claim about the world, relies on presuppositions that its own knowledge cannot establish. It remains in our power to reflect on those presuppositions and to compare them, and the way of life with which they are associated, with the presuppositions of other ways of knowing and living.

Arguments against positivism in social science in the 1950s and 1960s elaborated the claim that what we call 'mental' is a world of *shared* linguistic, cultural and historical forms of

life.[30] By contrast, the new neuroscience and materialist philosophy of mind has overwhelmingly taken it for granted that
mind is an attribute of the brains of *individual* people (or
animals). But mental states or functions, though (of course!)
in some way materially caused, psychological and individual,
are intrinsically social. This was the point G. H. Mead, for
example, made when he criticised the direction physiological
and experimental psychology was taking early in the twentieth century. He certainly had no quarrel with the advance of
natural science and he accepted a biological outlook on human
nature. Yet, he wrote, we must believe 'that human nature is
endowed with and organized by social instincts and impulses;
that the consciousness of meaning has arisen through social
intercommunication; and finally that the *ego*, the self, that is
implied in every act, in every volition . . . must exist in a social
consciousness with which the *socii*, the other selves, are as
immediately given as is the subject self'.[31] People identify and
characterise sensations, feelings or thoughts, however individual and subjective they may appear to be, as a social act: 'it
is possible to indicate in the nature of the consciousness which
psychology itself analyzes, the presupposition of social objects,
whose objective reality is a condition of the consciousness of
self'.[32] Even the way people experience mental events and talk
about them as internal states reflects historical and cultural
conventions. As Mead concluded: 'We may fairly say a social
group is an implication of the structure of the only consciousness that we know.'[33]

The point is of the greatest importance for any theory of
knowledge about being human. Natural scientists, when they
claim to understand people, frequently downplay or ignore it
completely; that is, they take for granted an untenable asocial,
individualistic view of what a person is. In doing so they

30 See Coulter, *Social Construction of Mind*, and *Mind in Action*;
   Geertz, 'Growth of culture', p. 76.
31 Mead, 'Social psychology as counterpart to physiological psychology', p. 97.
32 Mead, 'What social objects must psychology presuppose?', p. 109.
33 Mead, 'Social psychology as counterpart to physiological psychology', p. 103.

abstract the biological and psychological attributes of human nature from the social environment which gives them the meaning and nature that they have. Take perception, the example of the philosopher of science Marx Wartofsky: 'For the very foundation of what is distinctively human in perception is its character as a socially and historically achieved, and changing mode of human action; and thereby invested with a cognitive, affective and teleological character which exemplifies it as a social, and not merely a biological or neurophysiological activity.'[34] Any account of human nature or identity must include knowledge of the social form that nature or identity actually has, and this form has been reconstructed historically in the reflexive interaction of the form with language and culture.

It is very questionable to start with belief in autonomous individual reason as the basis for objective knowledge, in the manner in which Descartes proposed that he, himself, would doubt all that he could in his own mind so that he would be left only with what he could not doubt. Many neuroscientists, in this like common sense, imagine objective knowledge as the possession of an individual mind truly reflecting nature, like a mirror. Yet reasoning is a social achievement, and knowledge is a historical and collective phenomenon. As Martin Kusch critically concluded, 'what *is* individualistic . . . is to construe theories as the possessions of individual minds or brains'.[35]

Much is made of the continuity between animals and humans in modern science, and for good evolutionary reasons. Nonetheless, it is of considerable import to acknowledge that animals do not have historically constituted cultures in the way people do; this is why we, mostly, call animals 'animals'. It is also one reason why, for some purposes, it is possible to refer to the human sciences in contrast to the natural sciences. We do not, and cannot, define what makes a human being solely in terms of anatomy and physiology. 'The question

---

34 Wartofsky, 'Perception, representation, and the forms of action', p. 223.
35 Kusch, *Psychological Knowledge*, p. 317.

"What is a person"? . . . [is] a question for social decision rather than the discovery of a unique natural essence. A person is one whom we admit into our social group with the rights and duties of a responsible individual. The decision has varied from culture to culture, sometimes including and sometimes excluding such groups as slaves, criminals, lunatics, lepers, women, and even to some extent in our society dogs, cats, whales, and foxes.'[36] The fact that we sometimes exclude some people (like slaves) from the human realm and, conversely, admit some animals (like pets) into it shows rather vividly the social nature of the category 'human'. In the materialist natural sciences humans are indeed, among other things, animals. But in wider cultural life we draw distinctions. And one of those distinctions concerns the difference between explanation in causal material terms and understanding what action, and the life of mind, means for a socially located person. It is possible cultural life will change, as I imagined earlier, and the distinction will disappear; but if this happens it will be as a result of a change in cultural life and not because empirical knowledge has disproved the distinction.

Consider this example. Computer-generated pictures of the brain produced by magnetic resonance imaging (MRI) or by positron emission tomography (PET) have become extremely familiar.[37] These images are striking and reproducible; their appeal, however, also comes from commentators who say that they make a picture of previously invisible internal processes like feeling and thought. Two points tend to get lost. Firstly, the scientists and engineers who produce the images can choose, within the extremely wide range the existing technology permits, to give the images any form and colour they wish. The images chosen are a product of the scientific and engineering culture. Secondly, while the images show slices of changing patterns of events taking place in different parts of the brain, they do not show brains thinking or feeling. The brains shown are in people, and it is *people*, who are who they are in a

36 Arbib and Hesse, *Construction of Reality*, p. 83; see Putnam, *Reason, Truth and History*.
37 For an ethnography of the field, Dumit, *Picturing Personhood*.

social and historical context, who think and feel. This was clear long before the advent of such technology. As G. H. Lewes observed in the midst of the Victorian debate about materialism: 'It is not the brain which feels and thinks, it is the man.'[38] However much such knowledge develops, the human sciences, which try to understand human beings not brains, are left nearly where they were before. (This argument does not detract from the value imaging has in helping scientists differentiate and locate brain states and doctors in using such information as a sign of illness.)

Knowledge of the differential activity of the brain during different kinds of mental activity is one kind of knowledge. In itself, however, it says nothing about either what causes mental activity or what that activity is about. Different knowledge is needed to answer such questions, and in the latter case this certainly requires knowledge of language. It may also well be that we will not understand the causes of consciousness independently of knowledge of the causes of language and hence of people as social beings; at least some neuroscientists, including Steven Rose and Wolf Singer, have taken this view.[39] The argument leads towards the conclusion that there are different kinds of knowledge, different kinds of science – different kinds of rational understanding. There are human sciences as well as natural sciences.

### Does reflexivity demarcate the human sciences from the natural sciences?

Behind the claim that there are different forms of understanding in the natural and human sciences lies a claim about the distinctiveness of human action. This distinctiveness is intimately bound up with the reflexive phenomenon, the capacity of thought or feeling to be thought or feeling of itself. Not surprisingly, a number of writers have assumed that reflexive knowledge differentiates the human from the

---

38 Lewes, 'Spiritualism and materialism', p. 712. Also Midgley, *Beast and Man*, pp. 98–9.

39 Rose (ed.), *From Brains to Consciousness?*, pp. 14–15, 241–5.

natural sciences. But what sort of claim is this, and can it be sustained?[40]

Humanistic authors, like Cassirer and Collingwood, identified the capacity for reflection, and for the re-creation of the reflective subject through reflection, as a unique human attribute and hence as the distinctive content of the sciences of being human. By means of language and consciousness, they claimed, humans together, and only humans, have engaged in a never-ending process of reflecting on what they know about the world and, in the process, of changing themselves. Collingwood wrote:

> If that which we come to understand better is something other than ourselves, for example, the chemical properties of matter, our improved understanding of it in no way improves the thing itself. If, on the other hand, that which we understand better is our own understanding, an improvement in *that* science is an improvement not only in its subject but in its object also ... Hence the historical development of the science of human nature entails an historical development in human nature itself.[41]

If this is so, it appears possible to define the human sciences as the disciplines concerned with the reflexive process. It may well be that higher apes do, to a degree, consciously reflect; but, if they do, human beings still have a historical culture, which is re-creating itself (with, as it happens, devastating effects on the apes), which apes do not. As Michael Billig observed: 'Bees might be able to wiggle their stomachs in such a way as to communicate where pollen is to be found, but neither bees, nor instructed chimpanzees, can reflexively use words to argue about the meaning of words.'[42]

What is at stake is no neutral logical exercise. From Aristotle to Gadamer, many people have believed that this demarcation of the human from the natural domain is bound up with the possibility of there being an ethical sphere. 'Human civilization differs essentially from nature in that it is not simply a

40 See Smith, 'Does reflexivity separate the human sciences?'
41 Collingwood, *Idea of History*, pp. 83–4.
42 Billig, 'Psychology, rhetoric and cognition', p. 126.

place where capacities and powers work themselves out . . .
Thus Aristotle sees ethos as differing from physis in being a
sphere in which the laws of nature do not operate, yet not a
sphere of lawlessness but of human institutions and human
modes of behavior which are mutable.'[43] The being of being
human has become what it is through reflexive action, the
argument goes, realising an *ethos*; nature, by contrast, is
the realm of *physis*. Evidently, how we classify knowledge is
not separable from a decision about the ethical content of
human life.

Does reflexivity therefore demarcate the human sciences as
knowledge of the realm of *ethos*? It is fairly clear that, in so far
as this question concerns a description of scholarly practice,
the answer is yes. But the answer, insofar as it is a philosophical question about being human, is not nearly so clear
cut. Indeed, the previous discussion of reflexivity suggests a
negative answer, in spite of statements like Collingwood's and
Gadamer's.

To see this, it is helpful to make some observations about
theory and practice in the activity of disciplines. A discipline
is by its nature a social institution where particular premises
and practices become embedded over time. Indeed, in time,
the consolidation of ways of thought gives the appearance of
naturalness to a field's knowledge-creating activity, with the
result, at the extreme, that anyone who questions the disciplinary practice may be accused of questioning objectivity and
rationality itself. But, as previously stressed, it is, in principle,
always possible to question the premises of a discipline. In
practice, however, there are different conventions in different
settings about examining the in-built presuppositions both of
particular disciplines and of science in general. At one end of
the spectrum of practice are philosophers and writers on literary and cultural theory, whose discipline it is to build reflexive
critique, and a demonstration that closure of meaning is never
complete, into every statement. At the opposite end of the
spectrum, there are natural scientists who know that a stone

43 Gadamer, *Truth and Method*, p. 312.

is a stone, or a gene a gene, and whose business it is to close meaning about the nature of these things and least of all to engage in reflexive analysis of the history and culture that named such things in the first place.

Understood as a matter of practice, reflexivity clearly does separate the human and natural sciences: many people in the human sciences, influenced by the humanities, have begun to practise disciplined reflexivity, while many natural scientists think there are much more interesting things to do. This may appear a mundane conclusion. But it makes it possible to be clearer about a point and a question which are far from mundane. The separation between the natural and human sciences is a matter of convention, not principle, since, as analytic philosophers have shown, all bodies of knowledge contain premises in principle open to critique. And is it indeed the case that knowledge of humans changes the object of knowledge while knowledge of the physical world does not? It is not obviously true. If one set of concepts rather than another frames knowledge of the physical world, then it is in principle possible to change the framing concepts. And, except perhaps for people who believe they have access to a noumenal world beyond the reach of language, this amounts to the claim that the world changes along with our knowledge of it. There is a substantive sense in which changing concepts changes the world.

Stated baldly, the claim that the world changes along with our knowledge of it is a claim that may cause both ordinary people and natural scientists to bridle. It offends against a common, and common-sense, distinction between observer-independent properties and observer-relative properties. Searle, for example, has taken this distinction to be fundamental: 'That an object has a certain mass is an intrinsic feature of the object. If we all died, it would still have that mass. But that the same object is a bathtub is not an intrinsic feature; it exists only relative to uses and observers who assign the function of a bathtub to it.'[44] What can one say shortly in response,

---

44 Searle, *Rediscovery of the Mind*, p. xiii.

without getting entangled in deep-lying questions about epistemology and metaphysics?[45]

Firstly, there are philosophers, for example Hilary Putnam, who have rejected the reality/appearance dichotomy because it 'presupposes what Kant called "the transcendental illusion" – that empirical science describes (and *exhaustively* describes) a concept-independent, perspective-independent "reality"'.[46] The observer-dependent and observer-independent dichotomy breaks down. In Putnam's persuasive view, there are other forms of reason than natural science, there are ethical facts as well as material facts and hence there is no one neutral stance from which to view 'reality' independently of the particular purposes in regard to human flourishing to which we are, as it happens, committed. Secondly, it is open to question whether it makes sense to say that mass (or any other supposedly observer-independent property) continues if there are no people to use the word in a meaningful way. Stuff, we may suppose, endures, but it has no nameable properties unless people so name them. In this connection it is important to keep in mind 'that the social component of scientific knowledge is *not* to be set against the causal role of unverbalized natural reality: the social component is seen as a condition for having experience of a recognized kind and for representing that experience in linguistic form'.[47] Thirdly, there are systems of metaphysics – Whitehead's organicism, for example – which do not separate the world and the knower, and in these systems of metaphysics change in the latter is not divorced from change in the former. Indeed, much of pragmatism in philosophy was an attempt to rethink the taken-for-granted separation of the knowing process from its object. So was phenomenology. So, in a new idiom, is the philosophy of Latour, which adopted a 'radical realism', which does not

---

45 See Hacking's compromise position, 'that we "make up people" in a stronger sense than we "make up" the world', in *Historical Ontology*, p. 40.
46 Putnam, 'Place of facts in a world of values', p. 162.
47 Shapin, 'How to be antiscientific', p. 102 note.

separate subject and object in the manner of common sense.[48]
Fourthly, while it may appear self-evident that a telescopic
observation of a galaxy does not change that galaxy, in the
case of many other observations the matter is not at all so
clear cut. Quantum physics, especially the principle of indeterminacy, has given rise to intensive debate on the question,
and, in a lay person's understanding, many scientists appear
to have concluded that what is observed changes along with
observation of it at the sub-atomic level. In the area of the
human sciences, for example, in knowledge of the economic
market, there would appear to be no grounds for holding that
the object of knowledge exists independently of our accounts
of it. Across this continuum of phenomena, where should we
draw the line between observer-independent and observer-
dependent properties? But it is surely a fifth point which has
most weight and obviousness: knowledge as practice, as technology, manifestly changes the world.

Technology is material reflexivity, as the previous chapter
argued. There is a remaking of the world through reshaping
the environment, cyberspace, virtual realities and re-engineered
or even new forms of life, paralleled by cultures of altered
consciousness and new expressions of identity. We are certainly
altering both human identity and the identity of the world in
which people live. Attempts to draw a line between pure and
applied knowledge have collapsed. Collingwood was simply
wrong to write that 'nature stays put, and is the same whether
we understand it or not'.[49] If telescopic observation of galaxies
does not alter those galaxies, in the ordinary sense of the verb
'to alter', this is a matter of contingent limitations of power
not of theoretical principle.

The conclusion appears to follow: the reflexive implications
of knowledge do not separate knowledge in the human sphere
from knowledge of physical nature. Reflexivity does not
demarcate the human sciences. There is, however, a possibly
important qualification to consider.

48 Latour, *Pandora's Hope*.
49 Collingwood, *Idea of History*, p. 84.

Behind statements like Collingwood's and Gadamer's, and like those of Hampshire, Taylor and Winch cited earlier, that knowledge of humans changes humans in a way knowledge of the physical world does not change the physical world, is a normative premise: the word 'human' marks out an ethical domain. It is a premise in tune with humanistic intuitions, 'that ability to recognize universal – or almost universal – values, [which] enters into our analysis of such fundamental concepts as "man"'.[50] It is also in tune with the kind of philosophical anthropology which defined 'the human' in terms of the reflexive act, in which a being comes to be not only self-interpreting but self-constituting. Epistemological argument suggests that there is an important sense in which reflexivity is a potential dimension of the knowing act itself and hence cannot differentiate the human from the natural sciences. But ontological argument returns to the human being as the agent of the reflexive act and, in doing this, presupposes the distinctiveness and priority of what is human. Behind the proposition that reflexivity demarcates the human sciences is something more than an empirical or factual claim that human beings exhibit a certain capacity and animals do not. This 'more' is a feature of what it is *to be* human.[51]

This is what Foucault so influentially perceived and criticised. His work leaves a legacy of suspicion about the humanistic underpinnings of any claim to the distinct nature of knowledge in the human sciences. And yet – and here we get caught again in the apparent paradoxes of reflexive argument – by virtue of the critique which he mounted, he engaged in the kind of reflexive move which the tradition he attacked took to mark out being human. While conjuring up a spectre, which has not gone away, that reflexivity is a feature of a particular historical truth, and not the means to eternal truth, Foucault practised a kind of reflexivity. To think through these matters further requires a discussion of historical knowledge and the historical constitution of being human.

50 Berlin, 'Does political theory still exist?', p. 27.
51 See Taylor, *Human Agency and Language*.

In so far as reflexivity refers to a dimension of what people do, it does appear possible to characterise the human sciences by their search to understand the reflexive process. The natural sciences do not have this task. But this is a matter of convention and not logic. It is certainly open to study how human activity has given rise to the natural sciences, and how these sciences have acquired such immense authority as objective knowledge that they, and only they, in some secular settings, do not have to engage with a reflexive examination of their presuppositions as a condition of exercising authority. This is the subject matter of the history, sociology and philosophy of the natural sciences, disciplines which, as they study what humans do, have claim to be human sciences.

The difference between natural science and human science knowledge is not that nature remains as it is as a result of knowledge and humans do not, or that humans are forces in the production of knowledge and stones are not, but that humans have reflected, used language and created culture, including the present kind of discussion of it. This is a humanistic presentation of the existence of special and, as far as we know, unique capacities in being human. At the centre of these capacities, however, is reflection, and it is the central insight, and difficulty, of the philosophy of the human sciences that this very reflection has turned on itself, questioning, and thereby altering, its own identity.

Comte intended his philosophy of science to exclude false knowledge; but he ended up excluding knowledge of the conditions of knowledge. There is more to the history of knowledge than lies on the path laid down by the natural sciences.

# 4

# Precedents for the human sciences

## *Geisteswissenschaft*

It is a commonplace to describe the humanities as interpretive and the natural sciences, including the social sciences insofar as they follow suit, as explanatory. The English language, but not those of continental Europe, marks the distinction by calling the latter sciences but the former not. The distinction is most questionable. There are differences between causally explaining physical events and understanding human action, but these differences do not demarcate the natural from the human sciences, or either from the humanities. The differences indicate the possibility of different modes of thought, for different purposes. Interpretation is not some kind of activity in addition to cognition; it is a feature of all cognition, however much there are different kinds of cognition.

A history of debate about the relations among the sciences (in the continental European sense) will support these arguments. It will suggest intellectual precedents for, and create a view of, what writers are now calling the human sciences. The history will strengthen links between the psychological and social sciences and the humanities disciplines. There is an impressive and pertinent literature, from Vico and Herder in the eighteenth century, through the moral sciences of the nineteenth century, to hermeneutic and anti-positivist writings a century later. In addition, historical knowledge suggests that the way scholars classify knowledge is constitutive of what they then think to be true about the world. Classification is not the dry exercise of putting things in pigeonholes but the act of creating the holes into which to put things.

In a text published in 1883, translated into English a century later as *Introduction to the Human Sciences*, Dilthey gave currency to (though he did not introduce) a distinction between *Naturwissenschaft* and *Geisteswissenschaft* which has coloured subsequent enquiry about the nature of the human sciences. Indeed, an over-simplified equation of the terms '*Geisteswissenschaften*' and 'human sciences' might suggest that Dilthey is a sort of founding father for the latter field. That is certainly not my suggestion. Not least, Dilthey was heir to debates about human self-understanding going back beyond the Enlightenment.[1] Moreover, it is questionable to translate '*Geisteswissenschaft*' by 'human science'.

Dilthey himself thought it odd that, even in German, there was no conventional term for the sciences where 'the human' is the subject, though he pinpointed the reason – the subject is, in a sense, everything, since it is implicated in all acts of knowing. The potential field of the human sciences is too broad for any one term, ever, to be satisfactory. Dilthey therefore thought that we must make do with terms which at least denote key features.

> The sum of intellectual facts which fall under the notion of science is usually divided into two groups, one marked by the name 'natural science'; for the other, oddly enough, there is no generally accepted designation. I subscribe to the linguistic usage of thinkers who call this other half of the intellectual world the 'sciences of the mind' [*Geisteswissenschaften*] . . . But the expression shares this drawback with every other one which has been current. Science of society (sociology), moral, historical, cultural sciences: all these descriptions suffer from the deficiency of being too narrow with respect to the object they are supposed to be expressing. And the name we have chosen here has at least this merit: it rightly identifies the central core of facts from which one sees the unity of these sciences in reality, maps out their extent, and draws up their boundaries vis-à-vis natural sciences, although imperfectly.[2]

1 See Ritter (ed.), 'Geisteswissenschaften'.
2 Dilthey, *Introduction to the Human Sciences*, pp. 78–9; also translator's introduction, pp. 31–2.

The term caught on as a focus for discussion, even though philosophers did not agree with where Dilthey drew the boundary with the natural sciences. The very reasons which made it so hard to find one term even in German explain why translators have not come to agreement either. The problem centres on the word '*Geist*', which, in romantic and idealist writing, is at times equivalent to 'the human spirit'. These words denote something which has collective as well as individual identity. 'Mind' is not an adequate English rendering of '*Geist*' because 'mind' has modern, secular, psychological and individualist connotations. (The idea of a collective mind sounds in English like a contradiction in terms.) The German word connotes purposive activity, in pursuit of truth, beauty and goodness, which the English word often does not. Moreover, if such activity is called 'spiritual' in English, it will lead many modern readers to think that the subject matter is religious. Translating '*Geisteswissenschaft*' as 'science of the human spirit' results in a phrase which is at best strange and, to Anglophone eyes, an oxymoron, a combination of contradictory ideas. In my view, this is indeed the most accurate translation; but because of what people will read into the word 'spirit' it is unusable. Guy Oakes, a translator and historian of the relevant literature, reasonably sometimes opted to translate the term as 'the sciences of the mind', which brings out in English the distinction which Dilthey and others drew between 'the human' and nature. But he also sometimes translated the term as 'the human sciences'.[3]

As a result of these problems, a number of other translators or commentators render '*Geisteswissenschaften*' as 'the human sciences' and imply it is simply a collective term for the psychological and social sciences, linguistics, anthropology and so forth.[4] This elides the late nineteenth-century and early twentieth-century German debate in which the word featured, a debate about the relationship between the new psychology and the social sciences and the existing disciplines of learning.

3 In Windelband, 'History and natural science', p. 173.
4 Ermarth, *Dilthey*, pp. 359–60; Burnham, 'Assessing historical research', p. 225.

Whether or not these sciences belong with the *Geisteswis-senschaften* was precisely the point at issue. To translate '*Geisteswissenschaft*' as 'human science' obscures this, however useful it is to have the collective term in English. Furthermore, Dilthey believed that the facts to which all the *Geisteswis-senschaften* contribute reveal the activity of rational mind, and the field as a whole therefore has an essential unity and philosophical coherence. Such unity and coherence, however, is certainly not the social reality, and many would think it not the goal, of the modern human sciences. A further problem is that, in other authors, notably Gadamer, both the original term '*Geisteswissenschaften*' and the translation 'the human sciences' denote what English-speakers would more commonly call 'the humanities', not psychology and the social sciences. Gadamer's discussion of *Geisteswissenschaft*, in the manner of earlier idealist writers, was about the distinctive human expression of the values of truth and beauty, and his usage restated the division of the study of what the mind achieves from the study of mind as a function or property of biological and social processes.[5]

Some modern writers also refer to 'the human sciences' because they want to bring the disciplined study of human subjects into connection with the humanities rather than the natural sciences; this was, for example, one thought behind the founding of the academic journal called *History of the Human Sciences*. Other translators have rendered '*die Geistes-wissenschaften*' as 'the cultural sciences', and this does have the advantage that it leaves the question where to place the psychological and social sciences open for debate. Habermas's translator noted: 'Since in the last analysis *Geist* refers to everything that exists by virtue of man's symbol-making capa-city, I have translated *Geisteswissenschaften* as "cultural sci-ences."'[6] This brings in its train its own problem, however, since Rickert, who wrote at length on these matters, used the term '*Kulturwissenschaft*', which is clearly translatable as

5 Gadamer, *Truth and Method*.
6 Jeremy J. Shapiro, note in Habermas, *Knowledge and Human In-terests*, p. 337.

'cultural science', in the context of an argument against Dilthey's term.[7]

If it is difficult to keep all these possibilities straight in one's head, it is not because of play with words. What at first appears a straightforward question is very far from it. This is perhaps the main point: how we divide up the world into bodies of knowledge is an act which different people carry out in different ways and for different reasons.

The concept of *Geisteswissenschaft* took root in Germany as part of a critical response to Mill's philosophy, recently translated in a scholarly edition when Dilthey wrote. Mill, in his *System of Logic*, under the heading of 'the moral sciences', argued that the inductive logic of knowledge in the natural sciences was the logic of all knowledge whatsoever, and he made his point by envisaging the future development of the sciences of political economy and of what he called ethology, the science of individual character.[8] The English word 'moral', in this context, shared a root with the word *'mores'*, or manners, morals and character; it connoted those dimensions of human life intrinsic to the nature of a being with mind or spirit. Thus Hume stated: 'By *moral* causes, I mean all circumstances, which are fitted to work on the mind as motives or reasons, and which render a peculiar set of manners habitual to us.'[9] A phrase such as 'the moral faculties' referred to a wide range of mental states, connected with action rather than abstract thought, but not necessarily moral judgement in the modern sense. Mill's usage retained some of this older meaning, but he put the term to use in a modern project which was to render people knowable like any other object in nature. His German critics, Dilthey among them, rejected this project. For them, as Gadamer noted, using the term the '"moral sciences", already indicates that these sciences make their object into something that necessarily belongs to the knower himself' and is hence not the same as an object in nature.[10] Thus

---

7  Rickert, *Kulturwissenschaft und Naturwissenschaft* (1902), ed. and trans. as *Science and History*.

8  Mill, *System of Logic*, Book VI.

9  Hume, 'Of natural characters', pp. 213–14.

10 Gadamer, *Truth and Method*, p. 552.

they thought that Mill misrepresented the concept of a moral science.

Dilthey belonged to a generation of philosophers which was at one and the same moment impressed and distressed by the claims made for knowledge in the physical sciences. These sciences appeared uniquely successful in establishing precise, rigorous and cumulative knowledge, and their authority had grown dramatically over the preceding century. In contrast, philosophy appeared imprecise and without direction. In short, the natural sciences, spectacularly, had made progress. One philosopher observed that some people believe 'that philosophy no longer exists, but only its history', since science has taken over the advancement of human understanding.[11] The popularity of writers who repackaged natural science as the answer to general questions of life heightened academic anxiety about the authority of the natural sciences. Meanwhile, Germany had become an industrial power, and there were shifts in wealth and status away from the elite to which academics belonged, and there was an extensive, and growing, organised working-class movement. Academics felt their values threatened and references to cultural crisis abounded. The fear was that the form of knowledge in the natural sciences, so successful in its own terms, could not by itself sustain, let alone provide a rational ground for, objective values. This, scholars thought, was the business of disciplines like philosophy, history and theology. There was therefore felt to be an urgent need to enhance the social status and philosophical authority of the humanities disciplines if the natural sciences were not in practice, even if not in intent, to undermine the basis of true culture and social order.

Many German-language scholars felt it to be a self-evident spiritual necessity to understand poetry and the arts, Greek civilisation, the growth of the modern nation state and the history of philosophy as the activity of the reasoning, feeling and willing mind or spirit. They decried a view of history as the operation of laws of physical nature, as proposed, for example, by H. T. Buckle. The intellectual challenge they faced,

11 Windelband, 'History and natural science', p. 170.

therefore, was to show how the disciplines of philosophy, history, philology, theology and the study of human culture and society do, as a matter of fact and not just aspiration, establish both rigorous knowledge and true values. In a number of disciplines, including psychology, the science of language and economic history, scholars were divided: some identified their field with the natural sciences, while others stressed the activity of the human spirit. In this situation, argument about cultural values became interwoven with the very identity of disciplines.

Another factor in the debates was the extraordinary intellectual and cultural status of history. A scholarly concern with historical facts had emerged as a hinge of criticism of Hegelian philosophy in the 1830s, and, in a number of respects, German historians took over from philosophy and religion the privilege of representing values in cultural life. Though practising a primary source based discipline, historians emphatically read values and purpose into history, thereby making historical knowledge a pillar of support for highbrow culture. As the historian J. G. Droysen wrote: 'History is the way in which humanity becomes and is conscious of itself. The epochs of history are . . . the stages of its self-knowledge, its knowledge of the world, its knowledge of God . . . History is humanity's awareness of itself, its self-consciousness.' Even the identity and purposes of a person, the 'I', is the subject matter of history: 'Historical enquiry presupposes the reflection that even the content of our ego is something mediated, something which has come to be, a historical result.'[12] These were grand, even grandiose, claims for the centrality of historical consciousness to knowledge and to the progress of the human spirit. Moreover, these claims turned out to be double-edged.

It began to dawn on Dilthey's generation in the last decades of the nineteenth century that there might be unpalatable implications in accepting that 'even the content of our ego is something mediated' by historical conditions. A little later, critics introduced the term 'historicism' to describe what they took to be a disastrous combination of features of historical

12 Quoted in Schnädelbach, *Philosophy in Germany*, pp. 34, 53.

scholarship: an interest in historical facts at the expense of meaning; exposition of the merely relative value of cultural phenomena; and the belief that 'all cultural phenomena are to be regarded, to be understood and to be explained as historical'.[13] The discipline of history, they believed, had in fact – in spite of its professed ideals – contributed to a crisis of reason and values by promoting both relativism and positivism. This was not an unreasonable view. Historical scholarship on the New Testament had produced the most profoundly unsettling intellectual arguments of the nineteenth century, potentially turning the central dogma of Christianity, the incarnation, into a matter of empirical evidence. At the end of the century, observers became distressed by the suspicion that if all human activity is historical in its nature, then reason itself is also historical. This outcome, with its relativistic implications, was in the view of most German scholars an abdication of the calling of reason and of the vocation of philosophy.

All the same, scholars had considerable difficulty in knowing how to respond, as the following discussion will show in a little more detail. Hegel's philosophy, which had put forward an answer by equating what is real in history with reason itself, had, in most quarters, long been discredited. Dilthey and a group of neo-Kantian scholars turned instead to the theory of knowledge in order to show that there is a supportive relationship between facts and values in humanity's knowledge of itself. Weber was intellectually and socially close to these philosophers but, as the remarks in his Munich lecture to students in 1919 signify, he painfully concluded that the philosophers had not succeeded, and would not succeed, in their self-appointed task. 'The fate of our age, with its characteristic rationalization and intellectualization and above all the disenchantment of the world, is that the ultimate, most sublime values have withdrawn from public life, either into the transcendental realm of mystical life or into the brotherhood of immediate personal relationships between individuals.' This is an age, he implied, of public knowledge but private values. How long what he and his fellow academics took to be

13 *Ibid.*, p. 36.

real values could last in such circumstances was a pressing and disturbing question. It was exacerbated, in Weber's argument, by his view that scientists do not have a special capacity or calling to articulate public values. For example, taking the question 'Does the end "justify" the means or not?', Weber replied: 'The teacher can demonstrate to you the necessity of the choice. More than that he cannot do, as long as he wishes to remain a teacher and not become a demagogue.'[14] This bit deep because it questioned both the self-assigned vocation of scholars like Dilthey and claims that the sciences like history give systematic form to knowledge of true public values. How values might be grounded, Weber in this lecture did not say, but he gave the impression that the task is one which reason cannot address. In comparison with the established sense of the calling of the German scholar, this was a voice of despair.

Over the next decade, the 1920s, the logical positivists formulated a programme to cut through all these tortured discussions by demarcating observation statements, about which one can claim veracity, from value judgements, about which one cannot. They found an audience, especially in the English-speaking world, where it was already commonplace to assert the logical separation of factual and evaluative statements, and to explain, rather than to validate, values by reference to feeling, intuition, custom or utility. It was a significant sign of this situation that the Anglo-American study of ethics became ensconced in universities as a sub-speciality of philosophy, even though philosophy itself was still sometimes rather misleadingly (given the earlier scope of the term) called 'moral philosophy'. The study of ethics and the empirical study of human activity proceeded as if they were separate fields. In Anglo-American academic life it became completely unacceptable for history or any other human science to claim to be a form of knowledge of true values – however much these sciences described or explained observed social values.

This is the context the social sciences or human sciences found themselves in at the time of, and even a century after, Dilthey and Weber.

14 Weber, 'Science as a vocation', pp. 30, 26.

## Vico and Herder

The pursuit of knowledge in the modern West was always part of a moral project, the reform of man's estate, in the seventeenth-century expression. Throughout their history, the social sciences have carried the imprint of their birth in the enlightened faith that knowledge will make people free. Indeed, the roots of both positivist social science and anti-positivist, interpretive theories of human action lie in early eighteenth-century writings. (I say nothing about and do not pre-judge the preceding centuries.) It was, in particular, Vico and Herder, working outside the main centres of self-conscious intellectual innovation in Britain and France, who laid the basis for modern distinctions between the natural and human sciences.

Vico's 'new science' originated in the late seventeenth-century debate over the relative moral, artistic and philosophical significance of the achievements of the ancient world in comparison with the new age. He effectively redefined the terms of this debate, and he did so while entirely rejecting Descartes's claim to have made a new start in philosophy. The debate had seen philologists, whose expertise was in re-creating knowledge from texts, set against natural philosophers, who turned, with Francis Bacon as a figurehead, to empirical knowledge of nature, or turned, in the manner of Descartes, to mathematical and deductive reason. Vico did not defend the ancients – he fiercely attacked the notion of an ancient wisdom – but he nevertheless defended texts, not nature or mathematics, as a source of truth. He had in mind, however, texts understood according to his new science, not as understood by earlier grammarians and philologists. This new science interpreted texts in the light of a historical consciousness which Vico, at least in part, derived from legal scholarship.[15] In any case, in Vico's hands, historical consciousness acquired status as a source of knowledge that it has subsequently never entirely lost. Vico's new science made historical knowledge the form of knowledge appropriate for understanding texts and, through texts, the spirit or purposive way of living of the

15 Kelley, *The Human Measure*, pp. 234–9.

authors of those texts. This made historical knowledge the key to knowledge of what it is to be human.

Vico's *Principii di una scienza nuova d'intorno alla natura delle nazioni* (1725, last revised 1744) sought truth through reading texts understood as the creation of the human spirit or mind at a particular stage of its history. Or, expressed another way round, it proposed to establish historical knowledge, and hence knowledge of the human mind, on the basis of the verbal and written record which earlier ages have left. It can be demonstrated through literature and mythology, Vico believed, that there is a universal pattern in the history of human activity. 'My New Science simultaneously offers a *history of the ideas, customs, and deeds of humankind.* From these three topics, we shall derive the principles of the history of human nature, which are the principles of universal human history that we previously seemed to lack.'[16]

As the historian Anthony Grafton commented, Vico's book 'is commonly recognized as one of the founding works of the modern human sciences'.[17] This is because Vico articulated several claims which reappeared in the writings of scholars who defended the autonomy of the human sciences. Firstly, the subject matter of the human sciences is the activity of human beings in the ways in which they are distinctively human. In the idiom in use from Vico to Gadamer, this subject matter is the human spirit. Knowledge in the human sciences is knowledge of the activity of mind in language, conduct, society and culture. Secondly, the human sciences have their own forms of knowledge, which centre on the capacity to frame purposes and act in the light of them. Philology, hermeneutics (as a practice rather than general theory of knowledge) and contextual approaches to meaning exemplify disciplined attempts to make the study of these capacities objective and rigorous. Thirdly, the human sciences and historical consciousness are intimately connected.

Vico is perhaps now most famous for the claim that a historical science is a source of knowledge in a way that a natural

16 Vico, *New Science*, p. 139 (§ 368).
17 Grafton, introduction to *ibid.*, p. xi.

science is not, since, as he held, we can know with certainty what we ourselves have made. The key passage is deservedly much quoted:

> Still, in the dense and dark night which envelops remotest anti-quity, there shines an eternal and inextinguishable light. It is a truth which cannot be doubted: *The civil world is certainly the creation of humankind*. And consequently, the principles of the civil world can and must be discovered *within the modifications of the human mind*. If we reflect on this, we can only wonder why all the philosophers have so earnestly pursued a knowledge of the world of nature, which only God can know as its creator, while they neglected to study the world of nations, or civil world, which people can in fact know because they created it . . . Because it is buried deep within the body, the human mind naturally tends to notice what is corporeal, and must make a great and laborious effort to understand itself, just as the eye sees all external objects, but needs a mirror to see itself.

This was a very challenging claim and the basis for what Vico intended to be 'new' in his science. This science, in opposition to Cartesian science, would describe and explain the histor-ical stages in human culture as the development of the human mind. In cultural activity, Vico argued, we find a 'mirror' of what we are; moreover, because that 'mirror' is self-created, what it reflects can be known with a certainty that cannot apply to what people have not created, namely, the natural world. 'For there can be no more certain history than that which is recounted by its creator.'[18] The roots of this principle lie in Christian mysticism. God 'Himself produces even the spiritual plan of His Creation, and the science of its coming into being. This is the train of thought that leads to the critical epistemolo-gical principle of the identity of *verum* and *factum* [truth and fact], to the *solus scire potest qui fecit* [only he can know who makes (the object)].'[19] As God created, and knows, so too peo-ple create, and know. Where Vico was radical was in thinking

18 *Ibid.*, pp. 119–20, 129 (§§ 331, 349).
19 Blumenberg, *Legitimacy of the Modern Age*, p. 285; also Berlin, 'Vico and Herder', pp. 30–41, 122–38.

that, because people have created their world historically, history can be a source of *certain* knowledge. (Whether, and how, he reconciled this with his commitment to Christian revelation is a separate question.) To know, in scholastic thought, was to know what something is, to know both what causes it to be and what ends it fulfils in being. Vico re-created this way of thought and argued that history records the causes – both the how and the what for – of expressive life. As Berlin wrote, scholars can re-create the world of other minds, and in this way 'knowing why, and not merely knowing that, or knowing how – is the nearest that man can attain to divine knowledge'.[20]

Vico, and later Herder and the romantics, held that to be human is to have the God-given gift of freedom. What humanity does with this freedom is give itself a history – a past, a present and a future. Physical nature knows nothing of this. 'When the heavens finally thundered, Jupiter created the world of men by waking in them the moral effort or *conatus*. The human civil world originated in this effort, which is proper to the mind and its freedom. By the same token, the world of nature originated in motion, which is proper to physical bodies, which are not free agents but are subject to necessity.' But, though freedom is proper to mind, the mind, no less than the physical world, Vico thought, acts according to natural law. It is therefore possible to establish a science of mind with universal scope. The history of the nations is the history of an underlying mental activity, however much an individual nation has developed in its own particular way. Mental activity was, Vico supposed, initially primitive, barely able to control itself and barely able to distinguish itself from the world of natural forces. The primitive mind was 'poetic': 'Human ideas sprang from the divine ideas formed by early people when they contemplated the heavens with the "eyes of the body", as we say, rather than the eyes of the mind.' Gradually, intelligence and civilisation replaced the poetry and heroism of Homer's world. 'Terrified by the aspect of the heavens, . . . [people] checked their urge to sexual intercourse, and instead subjected their

20 *Ibid.*, p. 48.

lustful impulses to a conscious effort. In this way they began to enjoy human liberty ... (Since the body is the source of desire, this liberty must come from the mind, and thus is proper to humankind.)'[21]

Vico's scholarship stemmed from Renaissance humanism, and he set out to correct the work of his predecessors and demonstrate in detail to scholars how they should interpret the poetry, myth and religion of the ancient world. In his own assessment, however, he had done much more than this and had demonstrated that philology is a science, a body of knowledge known to be true because derived from rational principles – the laws of the human mind. This argument vindicated literary scholarship against Descartes's contemptuous charge that Cicero's serving girl knew as much of Rome as any historian. Descartes created a hugely influential model of clear reasoning as the means to truth, and this manner of reasoning had its greatest successes in the mathematical science of physical motion. Vico, if with a great deal less clarity of expression and little contemporary influence, had a profound vision of a different form of knowledge for the human sphere. He found, in the capacity of mind to know itself, the basis for a science of expressive human activity rather than a science of quantifiable abstractions. He put forward history as the science of human nature.

Vico's work is an early modern precedent for a distinction between forms of knowledge based on texts, or culture, and forms of knowledge based on nature. This is the distinction now institutionalised in the separation between humanities and natural science faculties. It must be stressed that this is not a distinction based on what is and is not *science*. The great changes in natural philosophy in the seventeenth century, which culminated with Newton's work, did not restrict scientific knowledge to knowledge of the physical world, however much such knowledge gained in status and authority. Other sciences existed, including Vico's new science, the moral sciences, the *Geisteswissenschaften*, hermeneutics, *les sciences de l'homme* and, more recently, the human sciences.

21 Vico, *New Science*, pp. 311, 152, 483–4 (§§ 689, 391, 1098).

Herder, who worked as an official alongside Goethe in the duchy of Saxe-Weimer-Eisenach, learned of the work of Vico, but only after he had arrived, by a different route, at a similar way of thinking about human self-knowledge. Herder's work was to be much more directly influential. His apophthegm, 'we live in a world we ourselves create', was a formative influence on the German *Aufklärung* and its romantic aftermath. Berlin later used it as an epigraph.[22] It paved the way for the transformation of scholarship and the arts which elevated the self-reflecting subject to the central position in the theory of knowledge.

By the last quarter of the eighteenth century, a conception of 'the science of man' as a historical science was current in a number of areas of speculative philosophy, for example, on the origin of language. Herder developed this historical consciousness into a universal human science. For Herder, the very business of being enlightened is to understand how, over historical time, the different peoples of the world have formed and used language, through language created social life and in social life constructed the physical, economic, institutional, political and cultural world around them. By creating this human world and coming to a self-consciousness of it, Herder argued, as Vico before him, humanity brings to fruition God's purpose. Hegel transformed this belief into what he held was the realisation of the absolute spirit coming, in historical humanity and the nation state, to full consciousness of itself. The roots were religious, but in the hands of Marx this philosophy became a fully secular redemptive worldview: we can understand the social relations that people have brought into being for what they are, and hence we can act on the basis of true freedom not on the basis of necessity.

Herder composed his multi-volume *Ideen zur Philosophie der Geschichte der Menschheit* in the 1780s, some time after Kant had introduced him to advanced thought while he was a student in Königsberg. The book was a comprehensive history of human progress, making possible knowledge of human nature. He drew upon the tradition of teaching about the soul

22 Translated in Berlin, 'Vico and Herder', pp. 168, 229.

in German universities, the more recent interest in creating an anthropology, or empirical science of man, the investment in *Staatswissenschaft*, or science of the state, and also juris-prudence. While Herder was writing, developments in hermen-eutics and historical scholarship, especially at the University of Göttingen, were laying the basis for the more professional, self-consciously precise scholarship of the century that fol-lowed. Herder's own text, however, though full of historical detail, remained very eighteenth century in its scope and inspirational rhetoric. In Herder's understanding, the Deity had created men and women with mind or spirit, and this spirit's activity, its expressive power voiced in language and the sociability that language has made possible, leads to progress in science, technology, religion, the arts and political society. He inspired readers with the view that a power for progress, rather than sinfulness, lies within the human nature variously shaped in each individual.

Language itself, Herder supposed, originated with the spon-taneous expression of the spirit reflecting on feeling and experience: 'this medium of our self-feeling and mental con-sciousness is – *language*'.[23] Owing to differences of climate, geography, mode of subsistence and history, he concluded, each group of people has created its world in a somewhat different way, and hence each people forms a unique entity with its unique culture, a *Volk*. Language and culture *express* the historically formed, unique life of a people. As a con-sequence, to have knowledge of other people requires us, in the manner in which we know ourselves, to enter into a way of life and to understand the use and expressive life of language. There is no universal language, but language is a universal necessity. In Wilhelm von Humboldt's words, influenced by Herder, 'the *bringing-forth of language* is an *inner need* of human beings, not merely an external necessity for maintaining communal intercourse, but a thing lying in their own nature, indispensable for the development of their mental powers'.[24]

23 Herder, 'On cognition and sensation of the human soul' (1778), p. 211.
24 Humboldt, *On Language*, p. 27. See Taylor, 'Importance of Herder'.

Expressive language makes possible relations differing from those which result from communication involving only designative signs.

Herder turned away from philosophy, understood as a logical discipline, in order to write, under the heading of anthropology, about the knowledge which the senses provide.[25] His conception of this knowledge was profoundly historical: if we talk about human nature, we talk about a nature which has created itself. History, therefore, in his understanding, is the substance of what people are by nature not a record of what people have done with their nature in history. He turned the philosophy of history into a story with universal reach about the coming into being of the form of life able to comprehend its own coming into being. In reaction against parochial and emotionally constricted social conditions, Herder, and even more his readers, opened up an imagination in which the historical process of self-creation stands in for the Christian idea of a personal and redemptive God. (Herder himself, however, it should be remembered, as an ordained Lutheran, strongly maintained the idea of providence.) It is in the power of the soul to reflect on itself, and in this reflection humanity discovers itself: 'it is ascribed to the human mind in an especially splendid analogy with the mind of the deity that only *the mind of the human being knows what is in the human being*, so to speak, rests on itself and explores in its own depths'.[26]

Kant did not respond to Vico, but his critical philosophy put the arguments as to why it may be judged impossible to have the certainty Vico thought humanity, as its own creator, could have. The generation of idealists who followed Kant, however, notably Fichte and Schelling, thought it possible to have certain intellectual intuitions, and in their own intuitions they discovered the self-positing spirit. They laid out philosophical arguments for the romantic sensibility, the conviction that it is the inner 'lamp' of the human spirit, which shines

25  Zammito, *Kant, Herder, and the Birth of Anthropology*, pp. 148–60, 308–45.
26  Herder, 'On cognition and sensation of the human soul' (1778), p. 199.

most brightly in the artist, rather than the 'mirror' of an external world that makes knowledge possible.[27] Influenced by Herder, writers suffused this sensibility with a historical consciousness. Philosophically conceived as the self-reflection of spirit, it spread through scholarly activity in philology, history, philosophy, jurisprudence, theology, the arts and, indeed, natural philosophy, exactly in the period when these subject areas were taking firm disciplinary identity.

## The moral sciences

The universities were in a very mixed condition throughout the eighteenth century; there was much variation in what they taught and to what standard. As the contrast between Descartes and Vico shows, there were substantial differences of view about the foundations of a properly constituted science, and these differences entered into arguments about university curricula and the classification of the sciences. The medieval institution of the discipline of a systematically taught body of knowledge, as had existed for grammar, rhetoric, arithmetic, astronomy and music, for example, persisted. All the same, there was little agreement about the content and teaching of large areas of importance. Some institutions and scholars grouped fields of learning into natural philosophy and moral philosophy, but neither of these terms denoted an agreed, systematically ordered and unified discipline. Natural philosophy attracted much greater attention in the curriculum of late seventeenth-century universities than it had earlier, at the expense of traditional commentaries on Aristotle. The significance and independence of this field grew, but there were no separate disciplines of physics, chemistry, zoology and so forth – the modern natural sciences – until the nineteenth century.

'Moral philosophy' did not name a discipline but a cluster of subjects, including ethics, politics and rhetoric, relevant to the science and practical art of human conduct.[28] This was an

---

27 Abrams, *The Mirror and the Lamp*.
28 Schneewind, 'No discipline, no history'. See Smith, '"Nauka o cheloveke"'.

ill-defined field, though it was of self-evident importance to all three of the higher faculties of the university (theology, law and medicine), all of which also dealt with both the nature of being and the practical affairs of men and women. The most systematic study of human nature in earlier centuries took the form of commentaries on Aristotle's account of the soul, known as *De Anima*, and this study was often the culmination of training in the lower faculties. With criticism of Aristotle, and especially with criticism of his natural philosophy, this teaching became problematic, though it persisted in certain institutions through the eighteenth century. It was a sign of change that a number of seventeenth-century authors introduced a distinction between *psychologia* and *anthropologia* to describe studies respectively of the human soul and of the human physical constitution. This introduced into some classifications of knowledge the kind of distinction which Descartes made so emphatically in philosophy, the distinction between what pertains to mind and what to body. It also made the term 'anthropology' current.

In France, elements of Renaissance moral philosophy teaching, greatly enriched by debate over Descartes's work on the soul, the body and the control of the passions, continued into the eighteenth century. The *philosophes*, working outside and in opposition to the universities, transformed these enquiries into what they hoped would be a unified and enlightened *science de l'homme*. On the one hand, they drew studies of human life and of nature into a single whole, since both, in their view, are subject to natural law. On the other hand, their essays and books made a distinction between *l'homme physique* and *l'homme moral*, which separated the body from conduct and social relations as different objects of study. This distinction had medieval precedents, and it was used by pre-Cartesian writers of the French reformed church, but perhaps its spread, and changing meaning, was a sign of Descartes's influence. The word *'le moral'* proved useful, describing the activity and capacities of soul or mind in the (social) world, in contrast to physical conditions.[29] The *moral/physique* distinction was to

29 See Azouvi, 'Physique and moral', p. 270.

have a long institutional life and to give the human sciences a kind of identity in the French-speaking world which they did not have elsewhere.

Buffon's introductory discourse (1749) to his *Histoire naturelle* set out a programme for the study of *le physique*, while Montesquieu's *De l'esprit des lois* (1748), which was the single most influential book to demonstrate the order to be found among the chaos of human affairs, did the same for *le moral*. Later, in 1780, Condorcet used the term '*sciences morales*' to denote 'all those sciences which have as the subject of their researches either the human mind in itself, or the relations of men one to another'.[30] (Subsequently, he divided the category into *sciences psychologiques* and *sciences sociales*.) By this time, contemporary with Herder, study of *le moral* had taken a profoundly historical turn and become intertwined with speculative philosophies of history – 'the transformation of physical man and woman into moral man and woman'.[31]

The reorganisation of higher learning in France after the Revolution institutionalised the category of *les sciences morales*. For a few short but productive years, the Classe des Sciences Morales et Politiques, the second of the three sections of the new Institut National, became the home of the *idéologues*, the group of intellectuals whom Gusdorf made the culmination of his history of the roots of the human sciences.[32] The section was itself divided into *analyses des sensations et des idées*, *morale*, *sciences sociales et législation*, *économie politique*, *histoire* and *géographie*. ('*Morale*', in the feminine grammatical form in modern usage, denotes the subject matter of ethical judgement; but, historically, there appears to have been slippage between '*le moral*' and '*la morale*'.) Napoleon abolished the section in 1803, viewing it as a centre of intellectual opposition, and in its place created the Classe d'Histoire et de Littérature Ancienne. A later and more liberal government, however, under a decree of 1832, opened the Académie des

---

30 Quoted in Baker, *Condorcet*, p. 197.
31 Wokler, 'Anthropology and conjectural history', p. 34.
32 Gusdorf, *La Conscience révolutionnaire: Les Idéologues*, vol. 8 of *Les Sciences humaines*.

Sciences Morales et Politiques, restoring the name created in 1795 and giving the new academy standing beside academies in the natural sciences and in the arts and belles lettres.[33] This institutionalised a division between the natural sciences, the field of what I would now call the human sciences and letters and fine arts, and it contributed to the natural appearance which some kind of class of *les sciences de l'homme* continues to have in France. (The term '*les sciences humaines*' is not equivalent, because of a troubled modern administrative history in the area, and because Canguilhem, Foucault and others associated this term with uncritical acceptance of the subject matter of institutionalised fields of research.)

Montesquieu, Condorcet, the *idéologue* Destutt de Tracy and, later, Comte, whether writing from a position within established institutions or not, all endorsed a form of knowledge in the *sciences morales* with close kinship to knowledge in natural philosophy. Condorcet clearly stated the view which Comte and Mill later elaborated: 'In meditating on the nature of the moral sciences, one cannot indeed help seeing that, based like the physical sciences upon the observation of facts, they must follow the same method, acquire an equally exact and precise language, attain the same degree of certainty.'[34] Destutt de Tracy spelt out the theory of mind and cognition which made this form of knowledge appear the only one possible, and which hence justified the view that the physical and moral sciences have a common method, even if they have different subject matter. Comte re-created this unified theory of knowledge as the historical march towards positivism.

The most interesting and influential curricular developments in eighteenth-century Britain occurred in Scotland. The universities of Aberdeen, Edinburgh and Glasgow brought together different parts of seventeenth-century teaching under the heading of moral philosophy, made this a major part of education and appointed a succession of eminent men to chairs in the subject. The most famous of these professors, Adam Smith, lectured on jurisprudence, language, history and ethics, as

33 See Leterrier, *L'Institution des sciences morales.*
34 Quoted in Baker, *Condorcet*, p. 86.

well as on what he called 'the science of a statesman or legis-
lator', which included political economy.[35] Moral philosophy,
in this setting, approximated to the science of man, and it
covered teaching on many aspects of human nature and conduct, from the science of the soul to jurisprudence. As in
France, there was rich interplay between investigations in
natural philosophy and moral philosophy, reflecting belief
about the continuity of natural law across physical and human
domains. Natural philosophy inspired Hume, for example,
when he made his now famous declaration at the beginning
of his *Treatise of Human Nature* (1739–40): 'There is no question of importance, whose decision is not compriz'd in the
science of man.' His approach was, in his own word, 'experimental'; he intended to be empirical but nevertheless to arrive
at conclusions about the universal attributes of human nature.
In practice, he turned to four sciences which were not part of
natural philosophy – logic, morals, criticism and politics – for
an understanding of the human life. 'The sole end of logic is to
explain the principles and operations of our reasoning faculty,
and the nature of our ideas: morals and criticism regard our
tastes and sentiments: and politics consider men as united in
society, and dependent on each other.'[36] It is, Hume thought,
the business of logic to clarify what it is possible for us to
know on the basis of experience; morals and criticism examine how we judge what is good and beautiful on the basis of
our feelings; and politics studies how reason and feelings lead
us to civilised living. This was a blueprint for the British conception of the moral sciences as they later developed (though
Hume's passage was not itself a particular influence).

In the wake of Montesquieu's work, a significant group of
authors, which included Adam Ferguson and James Millar
as well as Smith in Scotland, transformed moral philosophy
into a historical field, a field which began to produce detailed
accounts of the material and moral conditions at each stage
of human development. Their subject was human nature, but
they conceived this in much broader social terms than became

35 Smith, *Wealth of Nations*, vol. 1, p. 428 (Book IV, Introduction,
§ 1).
36 Hume, *Treatise of Human Nature*, pp. xx, xix.

*Being human*

the norm in the later specialised biological, psychological, political and economic sciences. The eighteenth-century moral philosophers were concerned with human action as it contributes to the social good, rather than individual happiness or well-being, and with the regularities of human nature, rather than individual subjectivity. This was reflected in the use of the word 'moral' to denote the sphere of human character and conduct, a usage which built into the language the assumption that descriptive knowledge of human nature is at the same time prescriptive knowledge of what it is good for men and women to do. In the Christian natural law tradition in which moral philosophy developed, human nature is a moral nature. Hume's subsequently celebrated distinction between 'is' and 'ought' statements did not, in the eighteenth century, affect the development of moral philosophy, of which he, too, was an exponent. Moral philosophy teaching developed alongside teaching in natural philosophy, and they formed complementary fields of knowledge of the providentially ordered world.

Between about 1780 and 1830, the moral philosophy agenda began to split up, creating specialised fields taught as separate disciplines. The clearest and earliest case is that of political economy, which, by the early nineteenth century, was a markedly specialised science – in the eyes of one modern observer, the first human science.[37] This shift towards disciplinary specialisation perhaps lay behind the growing tendency to use 'the moral sciences' rather than 'moral philosophy' as a collective term. With the establishment of University College, London, in the 1820s and the later reforms at Oxford and Cambridge, the disciplines of the moral sciences became a standard part of English higher education. This happened just at the time when the disciplines of the natural sciences (as opposed to unified natural philosophy) began to acquire status in the national culture as the most authoritative source of knowledge. From the beginning, therefore, there was a question about what status the moral sciences have if they are thought distinct from, as opposed to integrated with, the natural sciences.

37 Manicas, *History and Philosophy of the Social Sciences*, p. 36.

On the ground, subjects like Classics, history, philosophy and literature took shape, even while they became extremely disciplined, as a form of moral education or cultivation of character, and this embedded in British life assumptions about the different purposes of the sciences (i.e. the natural sciences) and the arts or humanities (which included areas of the moral sciences). Study of what became the psychological and social sciences lay somewhere uncomfortably in the middle; quite how this middle ground should develop was the subject of repeated argument.

This had early expression in the clash between Mill and Whewell, who, in 1838, took up the Knightbridge Chair of Moral Philosophy in the University of Cambridge, over the logic or manner of securing knowledge in the moral sciences. Mill wrote: 'the science of Human Nature may be said to exist, in proportion as the approximate truths, which compose a practical knowledge of mankind, can be exhibited as corollaries from the universal laws of human nature on which they rest', and when he referred to 'universal laws' he clearly meant laws comparable with the physical laws of nature.[38] By contrast, Whewell's conception of the moral sciences built moral judgement as a regulative principle into the fabric of knowledge. This was then accepted practice in historical writing and in the kind of political economy to which Whewell himself contributed. But over subsequent decades both the natural sciences and the academic disciplines in the arts increasingly outlawed this position – at least, in principle. This left moral ideals as something for education and scholarship to sustain, part of the formation of moral character, but they were not part of the content of knowledge.

Mill included history, political economy and a future science of society in the moral sciences. It is significant that he excluded ethics, strictly understood, because he held ethics to be an art, the application of knowledge not a form of knowledge. This signalled the great distance between his utilitarian notion of what is good and both Whewell's Anglican sense of 'Duty' and the German scholars' sense of what is constitutive

---

38 Mill, *System of Logic*, p. 848.

of humankind's rational being. For the idealists, values are
real and hence the proper subject matter for science. For Mill,
and for the British utilitarians for whom he spoke, however,
values derive from what actions mean for the pleasure and
pain which come with human nature, and it is human nature
which is the proper subject matter of science.

Mill looked forward to the time when the science of character would make possible the art of right government. In this
context, his view of the moral sciences was distinctive. He
took the view that the science explaining how individual character forms (which he, and almost he alone, called 'ethology')
is the science on which knowledge of society turns. He thought
that the whole future of social science, and hence of a rational political order, depends on explaining how the individual
person acquires a particular character and how this character
influences social life. 'Mankind have not one universal character, but there exist universal laws of the Formation of Character. And since it is by these laws . . . that the whole of the
phenomena of human action and feeling are produced, it is
on these that every rational attempt to construct the science
of human nature in the concrete, and for practical purposes,
must proceed.'[39] But Mill was no necessitarian. While he explained human character, in the last resort, by historical and
social circumstances, he remained a Victorian moralist in his
views about how people contribute to their own circumstances.
As a powerful advocate of education and reform at all levels
in society, he came close in his practical recommendations to
German ideals of the education of the spirit. But there were
German critics who feared that the logic of his position led to
nihilism.

Mill, for his part, directed every aspect of his work against
what he regarded as the idealist philosophical camp in Britain, regarding idealism as a conservative brake on rational,
progressive ethics and politics. Whewell, whose long volumes
on the history and philosophy of the inductive sciences had
appeared not long before Mill's *Logic*, was one of his principal
targets. Whewell was certainly conservative in many respects;

39 *Ibid.*, pp. 864–5.

nevertheless, at Cambridge he led the movement for curricular reform that, after 1848, made possible teaching in history, ethics and political economy in a newly created moral science faculty. Mill and Whewell both wanted to encourage the moral sciences and both took it for granted that subjects like history and political economy were sciences. Where they differed sharply was over 'logic', how to arrive at knowledge. Whereas Mill promoted induction, Whewell believed that an *a priori* regulative principle shapes the form of knowledge in each science, and that there is a correspondence between these principles and the order of the God-created world. A science advances, he argued, by the creative interaction ('the Fundamental Antithesis') in the mind of the principle, or 'Idea', and sensory experience of facts.

While Whewell's books on the history and the philosophy of the inductive sciences did not include the moral sciences, he made it clear that in principle his arguments extended to them. Moreover, a large part of his concern with the right method in science appears to have originated with debate about the relation between political economy and moral character, and, more generally, about the contribution of science to moral culture.[40] The moral sciences – in which at one point he listed ethnology, glossology, political economy and psychology, and in which at Cambridge he installed history, politics and moral philosophy – differ from the natural sciences, Whewell thought, in the nature of their shaping 'Idea'. This is the 'Idea of Duty' or moral obligation. Our knowledge advances in the moral sciences, he maintained, in as far as this 'Idea' binds together ('colligates') the facts of human social life. 'In Morality, in Legislation, in National Polity, we have still to do with the opposition and combination of two Elements; – of Facts and Ideas; of History, and an Ideal Standard of Action; of actual character and position, and of the aims which are placed above the Actual . . . In these cases, indeed, the Ideal Element assumes a new form. It includes the Idea of Duty.'[41] For Whewell, therefore, what differentiates the physical and

40 Yeo, *Defining Science*, pp. 102–6, 116–44, 193–8.
41 Whewell, *Philosophy of the Inductive Sciences*, vol. 1, pp. x–xi.

moral sciences is human moral nature and the shaping of knowledge about the human sphere which follows from it. The moral sciences are intrinsically normative; historical knowledge, for example, is objective knowledge of progress towards the moral ideals of Whewell's own Anglican and Victorian communion.

This was the authentic voice of British conservatism, to which Mill was heartily opposed. All the same, though Mill, unlike Whewell, thought knowledge in the moral and physical sciences has the same form, he, like Whewell, looked to the formation of individual moral character as the key to progress. And, during the time when the discipline of history consolidated its institutional position, it was precisely its 'special work . . . to give an historical or inductive basis, in other words a basis of fact, to moral science'.[42] History and ethics began to separate only in the last quarter of the nineteenth century, and, when they did so, neither speciality was quite what it had been before as part of the moral sciences.

The Moral Sciences Faculty continued to exist at Cambridge, but as a name for philosophy. Each of its subject areas – history, political thought, economics and philosophy – changed into a self-consciously professional academic discipline with its own institutional home. The pursuit of specialised expertise encouraged demarcation between factual statement and evaluative judgement and between sites of pure and applied work. A generalist, normative label, 'the moral sciences', lost its appropriateness.[43] Within academic philosophy, moral philosophy, often renamed ethics, became a sub-discipline cut off by special training and by the fact-value distinction from the wider concerns with human nature, human purposes and moral progress which had once informed it. This isolation of different areas of scholarship as different disciplines, and the isolation of a concern for facts from judgements about what is good, caused German scholars to refer to a crisis of culture. To many Anglo-American scholars it looked like the natural outcome of the empirical ethos of common sense.

42 John Seeley (1863), quoted in Collini, Winch and Burrow, *That Noble Science of Politics*, p. 214.
43 See Collini, *Public Moralists*, Part III.

Mill's work, appearing just when natural scientists were consolidating their institutional position and rapidly expanding the explanatory power and range of their activity, laid down a challenge to those who believed that human beings lie outside the scope of the natural sciences. German scholars, who took it for granted that the subject matter of the moral sciences is spiritual life, could not simply dismiss Mill with contempt in the way they did those they thought vulgar scientific materialists, like the physiologist Karl Vogt or the historian Buckle. They believed that Mill's empiricist logic and utilitarian approach to values made him blind to real truths about being human, but the discipline of his arguments forced them to say why. Gadamer thought that 'all the arduous work of decades that Dilthey devoted to laying the foundations of the human sciences was a constant debate with the logical demand that Mill's famous last chapter made on the human sciences'.[44]

The difficulties which the natural sciences created were evident in the lecture on relations between the sciences which the physicist and physiologist Hermann Helmholtz gave in Heidelberg in 1862. His starting point was the split between academics in the natural sciences and moral sciences (theology, law, politics, languages, art and history). He characterised the split in terms which were to remain topical: 'There is no denying that while the moral sciences deal directly with the most precious interests of the human mind and with the institutions it has brought into being, the natural sciences are concerned with external, indifferent matter, obviously indispensable because of the uses to which it may be put but apparently without any immediate bearing on the cultivation of the intellect.' Like Mill, Helmholtz hoped that the moral sciences would make a greater practical contribution to progress, but, unlike Mill, he did not think they could, for the most part, achieve inductive knowledge in the manner of the natural sciences. Rather, he held, 'the moral sciences deal with a richer subject matter' and 'have to do with judgments arrived at by psychological instinct . . . All this affords scope

44 Gadamer, *Truth and Method*, pp. 6–7.

for aesthetic but not for strictly logical induction.'[45] Here he gave voice to something many scholars feared: the identification of natural science with reason and the humanities with 'psychological instinct'. This was an intellectual crisis. Rescuing the moral sciences from the reputation that they rely on 'psychological instinct' and not 'reason' was the task which Dilthey and later Gadamer undertook. We may note, nevertheless, that though Helmholtz placed the natural sciences above the moral sciences in their use of rigorous argument, he did not question the reality and significance of what the latter study.

## Hermeneutics and 'understanding'

The transformation of the German universities in the early nineteenth century was a major intellectual and cultural event. The University of Berlin, founded in 1809–10, was both symbol and model: symbol of the importance of learning to German identity, and model for the reorganisation of scholarship along disciplinary lines.[46] The changes turned what had primarily been academies for future state servants into research universities. The outcome was a set of recognisably modern disciplines, each with a career structure in specialised research and teaching, designed to advance knowledge and strengthen the value accorded to learning. The significant reforms took place in what had in effect become the fourth faculty of the university, philosophy, and here scholars created the modern disciplines of the natural sciences and humanities, including what became philosophy in a specialised sense, and each discipline aspired to the status of science. Continued support for institutional expansion came from state officials, themselves educated in the universities and sharing their cultural ideals.

The achievements of scholarship, and increasingly the authority of scholars, rested on specialist knowledge and the

---

45 Helmholtz, 'Relation of the natural sciences to science in general', pp. 127, 137, 132.
46 Shaffer, 'Romantic philosophy and the organization of the disciplines'.

methodological rigour associated with it. While some people looked to philosophy to ground ideals and locate each discipline's knowledge within a unified understanding of reason, it remained a dream. Other people, like Helmholtz, thought philosophy a stagnant field, at least in contrast to dynamic disciplines like physics, physiology, philology and history. This situation, along with competition for funds and appointments, led to intense debate about the relations between subject areas and the significance of each for national culture. A conviction that the very nature and purposes of high culture were at stake informed debate about the classification of knowledge. Dilthey's work, as professor of philosophy in Berlin from 1882 to 1905, exemplifies this.

The authority of scientific scholarship in disciplines like history and philology rested on hermeneutics, reasoning which establishes the meaning of a text or other systematic expression of a symbol system. Originating in Judeo-Christian attention to the Word of God, it acquired the characteristics of a distinct area of scholarship in German Protestant exegesis of the Bible. Early scholars thought of hermeneutical work as primarily grammatical and logical, since their business was to sort out the correct expression of truth not to arrive at truth (which was already known). The reputation of the scholar and theologian F. D. Schleiermacher, working in the early years of the nineteenth century, is that he changed this, turning hermeneutics into a psychological and historical, rather than grammatical and logical, practice. That is, Schleiermacher re-created hermeneutics as the study of the relationship between the expressive life of the scholar and the expressive life of the scholar's subject, mediated by the text that one reads and the other created. Hermeneutics became a *historical* practice, a matter of interpreting the past, and it became a *critical* practice, a matter of arriving at a correct, or true, reading of what has been written down. This historical and critical view of reasoning applied to the record of the life of Christ transformed the Christian religion in the opening decades of the nineteenth century. At the same time, hermeneutics became the central epistemological and methodological preoccupation of the disciplines of jurisprudence, Classics, philosophy, history and the study of language. Subsequently, hermeneutics continued to be a set of specialised techniques, grammatical and logical, and historical and critical, for analysing texts. But it also became the covering term for discussion of interpretation, as opposed to explanation, in the theory of knowledge.

Dilthey's career began with writing Schleiermacher's intellectual biography, and it ended with the biography unfinished. In between lay extensive writings about what it means to interpret a life and about how, in general, to give interpretation scientific foundations. His starting point was the insight shaped in aphoristic form by the early nineteenth-century scientist of language, Humboldt: 'in order to understand each other, they [the interlocutors] must, in some other sense, have already understood each other'.[47] Hermeneutics, as a theory of knowledge, therefore attempted to comprehend what the reason of one person must be like in order to understand another person's reason. For Dilthey, historical understanding, to take the kind of knowledge with which he was most concerned, recreates the meaningfulness of a past world on the basis of what philosophical analysis has shown to be the underlying universals in life, universals expressed both in the knowing act and in the historical record. He initially hoped that psychology would become the science of these universals.

Expanding on the problems which biography presented, Dilthey became increasingly enmeshed in the philosophy of historical knowledge. He turned to psychology, understood as the science of the spirit or mind underlying history; and he turned to history for knowledge of the creative work of the mind and hence the source of psychological knowledge. The outcome was a deeply historical conception of being human and, and at the same time, of psychology as the science of the mind which transcends history. 'The human being as a fact which precedes history and society is a fiction of genetic explanation. The man whom sound analytic science studies is the individual as a component of society. The difficult problem which psychology has to solve is this: analytical knowledge

47 Humboldt, 'On the historian's task', p. 65. See Ermarth, *Dilthey*, p. 253.

of the universal characteristics of this man.'[48] The philosophical task was to elucidate the meaning and validity of these claims for history and psychology by showing them to be implied in the exercise of reason. Dilthey's project, however, remained incomplete; moreover, from the second half of the 1890s, reinforced by his reading of Husserl's attack on psychological solutions to logical and epistemological questions, he withdrew the hopes he had specifically placed in psychology.

Before this, Dilthey had established himself as a leading spokesman for the scientific authority of hermeneutic methods. It was in this context that he introduced his much-discussed distinction between *Naturwissenschaft* and *Geisteswissenschaft*. The sciences of the spirit, he thought, have an underlying unity and autonomy, which 'derives from the depth and fullness of human self-consciousness. Even when unaffected by investigations into the origins of the mind, a man finds in this self-consciousness a sovereignty of will, a responsibility for actions, a capacity for subordinating everything to thought . . . by these things he distinguishes himself from all of nature.'[49] His notion of the *Geisteswissenschaften* rested on a philosophical anthropology; it was not simply a collective term for the cluster of disciplines which formed the human sciences a century later.

There were a number of characterisations of the difference between knowledge in the natural sciences and in the sciences of mind in circulation. For Dilthey, the key distinction was that the natural sciences explain while the sciences of mind understand. This was more than a verbal refinement, since 'explanation' meant showing that a physical change is the outcome of deterministic, law-like events, and 'understanding' meant showing that actions are the outcome of rational purposes and values present in human consciousness. The human sciences, he wrote, 'have a basis and structure completely different from those of nature. Their object consists of given, not inferred, units which we understand from within;

48 Dilthey, *Introduction to the Human Sciences*, p. 94.
49 *Ibid.*, p. 79.

we know and understand here first, in order gradually to comprehend.'[50] Understanding (*Verstand*) claims knowledge about
a world structured by *meaning*, which explanation (*Eklärung*)
by itself cannot do.

Kant's authority for a distinction between reason (*Vernunft*)
and understanding was important. Kant attended in the critique of reason to the form that knowledge necessarily takes;
in the critique of judgement, he attended to the teleological,
purpose-oriented, form of knowledge, or understanding, specific to organic life and art. According to Dilthey, the scientific
understanding of human action shares this latter form. He
used the common German verb '*verstehen*' to signify this act
of understanding, which he initially represented in psychological terms and which he thought ubiquitous in human
awareness. Understanding is an act in which one living mind
knows another. 'I myself, who inwardly experience and know
myself, am a member of this social body and ... the other
members are like me in kind and therefore likewise comprehensible to me in their inner being. I understand [*verstehe*]
the life of society.'[51] Weber subsequently also made understanding central to his discussion of the way knowledge in the
social sciences must take account of beliefs as part of the
explanation of the human world. Weber, however, in contrast
to Dilthey (at one stage), did not represent the believing act in
overly straightforward psychological terms.

English-language translators often either leave '*verstehen*' in
German or have recourse to verbs like 'to grasp' or 'to enter
into' the meaning of a human action. It remained for analytic
philosophy (discussed in the previous chapter) to make more
precise the argument that human actions require explanation
in terms of intentions, reasons and purposes. All the same,
the language of 'grasping meaning' remained a mainstay of
informal accounts of knowledge in the humanities.

Philosophers of hermeneutics appreciated that interpretation involves circularity. Humboldt himself had drawn attention to this: 'All understanding presupposes in the person

---

50 *Ibid.*, p. 141.
51 *Ibid.*, p. 98.

who understands, as a condition of its possibility, an analogue of that which will actually be understood later: an original, antecedent congruity between subject and object.'[52] In consequence, it appeared impossible to validate it as the form of reasoning in the human sciences independently of establishing an anthropology, a philosophy of being human. As Heidegger was to note: 'In every case interpretation is grounded in *something we see in advance* – in a *fore-sight*.' And, as he went on to explain, we must accept that this circularity is given by the reflexive nature of being, and, therefore, 'what is decisive is not to get out of the circle but to come into it in the right way'.[53] The attempt to ground hermeneutics therefore opened into questions about the reflexive nature of knowledge. Thinking cannot avoid making presuppositions which, logically considered, it is the business of thinking to establish in the first place. Heidegger came to the conclusion that a new way of doing philosophy was needed and, in turning to a philosophy of being, believed that he stepped beyond the historical contingencies in the thought of philosophers since the Pre-Socratics. But Dilthey continued to discuss circularity as if it could be out-stepped, from within the theory of knowledge, by a kind of intuitive grasp of what is real in being human. For example, he called description 'an artistic process . . . intuitions determine the value of the results'.[54] It was just this kind of statement that led logical positivists in the 1920s, and many later Anglo-American empiricist philosophers, to dismiss such people as Dilthey out of hand. In their eyes, such statements confirmed that the humanities disciplines are not really sciences at all. But by no means all philosophers took this line. Gadamer, for example, aligned the sciences of being human with the arts, as Dilthey had done before him.

Dilthey was one of a number of scholars who thought that the problem of relations between the sciences came to a head in the case of psychology, an area of activity which, during his lifetime, took over the experimental methods of physiology

52 Humboldt, 'On the historian's task', p. 65.
53 Heidegger, *Being and Time*, pp. 191, 195.
54 Quoted in Ermarth, *Dilthey*, p. 252.

while still addressing philosophical problems. For much of his career, with his Schleiermacher biography in mind, he had hopes for a general theory relating cultural activity – in art, scholarship, politics and so on – to the laws of the expressive human mind. He conceived of a science of historical psychology which would link the rational activity of the individual mind to the rational creations of shared culture. This was an unexceptional position at the time: psychologists or scientists of language like H. Steinthal and Wundt, who wrote on *Völkerpsychologie*, articulated similar views. Dilthey, in writing biography, intended to relate the specific historical circumstances of an individual life to general knowledge of the relations of human nature with culture. But the task proved too difficult.

There was also an institutional dimension. Dilthey strove to influence the development of psychology as a discipline at a time when philosophy faculties were newly appointing a number of experimental psychologists. Not uncommonly, philosophers viewed the growing numbers of experimentalists, some of whom took positions previously occupied by traditional philosophers, as a sign of the wider crisis in values. Dilthey himself helped secure the appointment of Carl Stumpf rather than the narrowly experimental, though innovative, Hermann Ebbinghaus, to head Berlin's new institute for psychology in 1893. Stumpf was more acceptable to Dilthey because Stumpf, in a manner later developed by phenomenological psychologists, took the starting point of psychological analysis to be the qualitative phenomena of the conscious world. This appeared to allow scope for values in the subject matter of psychology, which Ebbinghaus's approach did not. In the midst of this academic struggle, Dilthey published a long study promoting history as the route to a psychological science of the value-expressing activity of human beings. 'Man does not apprehend what he is by musing over himself, nor by doing psychological experiments, but rather by history.'[55] Ebbinghaus responded in scathing terms, making it appear as

---

55 Dilthey, 'Ideas concerning a descriptive and analytic psychology', p. 63.

if there was a straightforwardly polarised conflict between treating psychology as a hard experimental natural science or as a soft humanistic subject.

Dilthey's project foundered, and he began to question the capacity of psychological knowledge to build the bridge between present minds and past minds. It appeared necessary, instead, to ground psychological knowledge, like all other forms of knowledge, in an analysis of reason and judgement. Dilthey was in the difficult position of having laid claim to the distinctiveness, and significance, of the sciences of mind precisely because they concern themselves with knowledge of absolute values, or the expression of the human spirit, while in practice the philosophical grounding of these values was in crisis. As he noted, in the philosophy of history 'the extremities of our consciousness – knowledge of reality and consciousness of value and rule – are tied together as one in their general conception'.[56] His project stumbled on the key question of the grounding of values. This made the whole claim to have distinguished *Geisteswissenschaft* from other knowledge vulnerable.

Dilthey and others understood Husserl to have provided a rigorous demonstration of the intrinsically evaluative structure of consciousness, and to have connected this with the theory of knowledge. It is the irreducible being of conscious experience, Dilthey believed, to present experience as having meaning and to be structured according to judgements of significance. 'The significance which a fact receives as a determinate link in the meaning of a whole is based on a vital relationship and not an intellectual one; it is not a question of imposing reason and thought on part of an event. We draw significance from life itself.'[57] From this vantage point, Dilthey returned to his concept of understanding. Knowledge in the human sciences, he concluded, involves understanding, and understanding is possible because of the common evaluative structure of consciousness. It is therefore for the sciences to bring together philosophical analysis of this structure with historical study of its expression in culture and conscious life.

---

56 Dilthey, *Introduction to the Human Sciences*, p. 134.
57 Dilthey, 'Construction of the historical world', p. 241.

What was felt to be a crisis of values concentrated minds on the classification of the sciences and the theory of knowledge. Dilthey's work was one point of reference; another was the rectorial address in Strasbourg (or Strasburg as it then was) of the philosopher Wilhelm Windelband in 1894. He introduced terms, which subsequently took on another life in analytic philosophy of history, and distinguished 'nomothetic' knowledge characteristic of (but not unique to) the natural sciences from the 'idiographic' knowledge characteristic of (but not unique to) the human sciences. His intention was to mark the differences between knowledge of causal activity according to natural law and knowledge of individual events. 'The purpose of these [natural science] disciplines is invariably the discovery of laws of phenomena. In contrast to these sciences, the majority of the disciplines that are usually called sciences of the mind have a distinctively different purpose: they provide a complete and exhaustive description of a single, more or less extensive process which is located within a unique, temporally defined domain of reality.' His point was not that the different sciences have different subject matter, but that they aim at different goals; the purpose of knowledge of nature is different from the purpose of knowledge of human values. 'In their quest for knowledge of reality, the empirical sciences either seek the general in the form of the law of nature or the particular in the form of the historically defined structure.' In the conclusion of his address, he briefly signalled the ultimate purpose which, he believed, lies behind the use of the idiographic form of understanding. Here he hinted at arguments going beyond epistemology to ontology. He upheld the irreducible value of the individual, whose presence is a unique historical event, and the 'incomprehensible character of the personality [which] emerges as the sense of the indeterminacy of our nature – in other words, individual freedom'. In contrast, he said, 'utterly indifferent to the past, the natural sciences drop anchor in the sea of being that is eternally the same'.[58] As his translator observed, Windelband's position derived from 'the Christian idea that values can be

58 Windelband, 'History and natural science', pp. 174, 175, 179, 184.

ascribed only to individual phenomena [, an idea which] has its origins in the conception of the Creation, the Fall, and the life of Christ as unique events endowed with unprecedented significance'.[59]

In Windelband's view, the search for general laws and the search for particular explanations demand different methods. Any particular science, however, may exhibit both nomothetic and idiographic elements. Accordingly, it is wrong (both Windelband and his contemporaries who criticised his work recognised this) to define the *Naturwissenschaften* as nomothetic and the *Geisteswissenschaften* as idiographic, even if it is thought that these terms specify the characteristic form of knowledge in each domain. Psychology, once again, presented a challenge – though it was the challenge to clarify its forms of understanding not to decide whether it is a science. Windelband himself thought that the aims of psychology were primarily nomothetic, though he stated that 'an empirical discipline as important as psychology cannot be classified unambiguously as a natural science or as a science of the mind'.[60]

Windelband's one-time student and colleague, Rickert, also stressed that what is at issue in the classification of the sciences is not subject matter but forms of knowledge. As he pointed out, there is no subject matter which cannot become the subject matter of the natural sciences: 'there is only one empirical reality'. Like Windelband, he accepted that *Geisteswissenschaft* differs from *Naturwissenschaft* because it (especially) requires methods appropriate for understanding individual entities – nations, persons, historical events, works of art – in their concrete particularity. This particularity in itself, however, is not what makes the difference. The sciences of the particular differ, fundamentally, because the interest in individual entities is an interest in, and expression of, the intrinsic and absolute value, the truth, goodness or beauty, that persons, bodies of knowledge, nations or works of art have. Rickert therefore preferred to refer to the *Kulturwissenschaften*, the

59 Oakes, *Weber and Rickert*, p. 43.
60 Windelband, 'History and natural science', p. 174.

cultural sciences, rather than to the *Geisteswissenschaften*, as he thought the latter's connotations tied the relevant sciences too specifically to an empirical science of mind. Like Windelband, but unlike Dilthey, Rickert classified psychology as primarily a natural science and viewed psychical life as part of nature. Rather than give priority to psychology, he wanted to understand politics, jurisprudence, political economy, theology, literature and art, as well as history, as the historically achieved coming into being, the culture, of particular realisations of what is true, good or beautiful. In contrast, he held, the natural sciences explain the law-like movements of physical nature, which have no meaning or value, and no history. He sharply demarcated nature and culture. 'Nature is the embodiment of whatever comes to pass of itself, of what is "born" and left to its own "growth." Culture, on the other hand, comprises whatever is either produced directly by man acting according to valued ends or ... whatever is at least *fostered* intentionally for the sake of the *values* attaching to it.' Values, Rickert argued, exist in the human achievement of knowledge about nature not in nature. 'We *can* also represent completely indifferent objects, altogether devoid of meaning, in their individuality if we *wish* to do so. In such cases, it is our act of will alone that makes this individuality "important" and thus endows it with value.'[61]

Rickert potentially laid the basis for understanding knowledge in the natural sciences as a cultural, not only cognitive, achievement. 'Not only are the natural sciences an historical product of civilized man, but also "nature" itself, in the logical or formal sense, is nothing but a theoretical value of *cultural* life, a valid, i.e., objectively valuable, *conception* of reality on the part of the human intellect.'[62] Such argument vindicated the cultural importance of the sciences of mind – they could not be threatened by the natural sciences – since it gave epistemological priority to historical knowledge of the human creation of knowledge over knowledge of nature. Yet Rickert

---

61  Rickert, *Science and History*, pp. 13, 18–19, 129–30. Also Rickert, *Limits of Concept Formation*.
62  Rickert, *Science and History*, p. 143.

was immovably opposed to the conclusion that scientific knowledge, of any kind, is relative. This marks his distance from
one influential strand of thought in the human sciences later
in the twentieth century. For a man of Rickert's philosophical,
social and cultural background, it was unthinkable to deny, in
the last analysis, the objectivity of what he tended to call real
values. If he concluded that nature is a historical category, he
nevertheless had faith that reason and truth are not.

Rickert and other scholars of his persuasion entwined themselves in circular argument, using history to reveal values and
presuming values in order to vindicate historical knowledge.
There was no escape without an explicit return to ontology,
perhaps of the Christian kind that Windelband hinted at in
his reference to 'individuals' at the conclusion of his rectorial
address. Cassirer later observed: 'If [the scholar] seeks to
establish . . . ["validity"] on the basis of history itself, he is in
danger of involving himself in a circular argument. If he seeks
an a priori construction of such a system, as Rickert himself
has done . . . it has been shown again and again that such a
construction cannot be carried through without metaphysical
assumptions, and that in the final analysis the problem ends
just where it began.'[63] Nietzsche had meanwhile shown how
persuasive the relativist perspective that Rickert and others
were so committed to denying could be, and Nietzsche's arguments were to prove hugely influential in the long run. All the
same, events in the twentieth century affected the confidence
in 'real values' more than any abstract argument. One cannot
mistake the declining power and importance of the 'mandarin' class, profession and culture to which these professors of
philosophy belonged.[64] As the Third Reich, subsequent political settlements in Europe and rise to power of so-called popular
culture showed, their values were indeed under threat.

Something of this failure, as it was felt to be, to secure the
basis of values entered into Weber's portrayal, in his manner
of life as well as in his writing, of the position of the moral
intellectual. Though he died in 1920, his attempt to clarify the

---

63 Cassirer, *Logic of the Humanities*, pp. 90–1.
64 Ringer, *Decline of the German Mandarins*.

nature of knowledge in the social sciences and his criticism of
scholars like Rickert lived on to haunt the English-speaking
as well as German-speaking world in the second half of the
century. The way Weber tackled questions about the nature
and objectivity of sociological concepts, the relation between
objectivity and values and the different forms of rationality in
social and natural science explanation shaped much later discussion. For this reason, he remained a figure of enormous
interest for the human sciences. What he did not do is reach
a solution for the crisis of values. Weber drew on Rickert's
work on values in the cultural sciences, and the difficulties,
and failures, of Rickert's search were also Weber's own.[65]

Weber stated clearly that purposes or problems, not subject
matter, separate the natural and the human sciences. 'It is not
the "actual" interconnections of "things" but the *conceptual*
interconnections of *problems* which define the scope of the
various sciences.'[66] He then linked this insight to his account
of the relation between values and objectivity in the sciences.
Unlike most of his German-speaking academic contemporaries
in the humanities and social sciences, and unlike students
embracing knowledge as a guide to life, Weber sharply differentiated empirical claims and normative statements. 'The
empirical disciplines', including the social sciences, 'are concerned only with the fact that the validity of a practical
imperative *as a norm* and the truth-value of an empirical proposition are absolutely heterogeneous in character. Any attempt
to treat these logically different types of propositions as identical only reduces the particular value of each of them.'[67] By
such firm statements he intended to defend the integrity and
objectivity of all the empirical sciences. At the same time,
he recognised that different sciences have different cognitive
goals and that the goals of the social sciences are prescriptive
and managerial as well as what he called 'formal' or purely
cognitive:

65 Oakes, *Weber and Rickert*.
66 Weber, ' "Objectivity" in social science', p. 68.
67 Weber, 'Meaning of "ethical neutrality"', p. 12.

There is no absolutely 'objective' scientific analysis of culture . . .
of 'social phenomena' independent of special and 'one-sided'
viewpoints according to which – expressly or tacitly, consciously
or unconsciously – they are selected, analyzed and organized
for expository purposes. The reasons for this lie in the character
of the cognitive goal for all research in social science which
seeks to transcend the purely *formal* treatment of the legal or
conventional norms regulating social life.[68]

Scientists who research culture have a normative interest which
differs from the 'formal' interest of natural scientists in general laws of nature. Social scientists select material in the
light of their evaluation of what matters, concentrating, for
example, on how particular beliefs support a particular way
of life, as Weber himself did in his study of the Protestant
ethic. This Weber thought necessary and proper: 'the problems of the social sciences are selected by the value-relevance
of the phenomena treated'.[69] But he thought it unnecessary
and improper for social scientists to claim that what they
select reveals an 'absolutely objective' stance. Rickert, by contrast, did think it the business of the academic to guide cultural
life in general, and students in particular, towards objective
values.

### Hermeneutics reconsidered: Gadamer

These debates about explanation and understanding as forms
of knowledge reappeared, though this time in the idiom of
English-language philosophy, in the 1960s and 1970s. German
philosophy of course did not stand still in the previous
half-century. In the 1920s, there was an acute consciousness
that the question of values, the question of the relationship
between intuited true values and reason, remained unsolved.
For all their insistence that knowledge in the sciences of the
human subject must put true values in their proper place,
Dilthey and like-minded scholars had not shown authorit-
atively how this could be done. Weber implied that the task

68 Weber, '"Objectivity" in social science', p. 72.
69 Weber, 'Meaning of "ethical neutrality"', p. 21.

was hopeless. Other philosophers, Husserl, with phenomenology, and later Heidegger, turned to what they regarded as the
more fundamental task: analysis of what can be said about
existence. Others thought it urgent to bring the human and
biological sciences together in a comprehensive philosophical
anthropology. Among these, Cassirer, who was one of Dilthey's
and Rickert's closest readers, is the philosopher who became
best known in the Anglophone world in the middle years of
the twentieth century. It was, however, Gadamer, who had
been Heidegger's student in the early 1920s, who re-examined
the tradition of hermeneutics and presented it in the form
which was most discussed in the last decades of the twentieth
century.

Cassirer, working in the shadow of Kant, approached the
problem of knowledge through the logical elucidation of the
way reason gives 'form' to experience. Reason has a unity,
and so far as the 'fundamental theoretical activities of the
human mind are concerned we can make no discrimination
between the different fields of knowledge'. However, there are
different modes of forming knowledge in different fields, differences between the field of physical causality and the field
of human freedom. The difference between these fields is a
matter of how we know, not a matter of what exists. 'Freedom and causality are not to be considered as different or
opposed metaphysical forces; they are simply different modes
of judgment.'[70] He accepted the descriptive value of the terms
'nomothetic' and 'idiographic' but thought it wrong to use the
terms as categories in the classification of knowledge, since
both types of knowledge are present in many sciences.[71] His
approach to the relation between the different sciences was,
therefore, historically and philosophically to analyse the 'different modes of judgment' in the different areas of life. If at
times there has been opposition between natural science explanation and cultural explanation, Cassirer argued, this is
because different forms of knowledge have different instrumental value and the purposes for which different forms were

70 Cassirer, *Essay on Man*, pp. 176, 193.
71 Cassirer, *Logic of the Humanities*, p. 120.

created have become confused. As a telling example, he cited the work of the nineteenth-century French historian and literary scholar, Hippolyte Taine. Taine, battling against a highly conservative literary establishment, rhetorically emphasised the need for literary scholarship to become a science comparable with a natural science. Yet, when he came to describe the content of literature and the other arts, he richly evoked aesthetic and moral meaning and did not restrict himself to causal, natural science explanation. Taine, in practice contradicting his own rhetoric, made a place for both natural science and humanist forms of reasoning. Cassirer concluded: 'Form-analysis and causal analysis are now seen to be orientations which, instead of contending with each other, complement each other, and which necessarily require each other in all knowledge.'[72]

When Cassirer referred to 'form-analysis' he had in mind the distinctive preoccupation of the humanities disciplines with the meaning that human events and actions have, and with the way acts of symbolisation constitute this meaning. By contrast, the natural sciences have a distinctive preoccupation with 'causal analysis'. This distinction served, in a general way, clearly to separate the natural sciences and the human sciences; a physicist analysing light waves and a literary scholar analysing a poem obviously seek different kinds of knowledge. Deploying philosophical analysis, however, Cassirer argued, it can be seen that both tasks involve form-giving, rational activity, as indeed does even the simplest statement about a fact. This form-giving 'shapes' – it creates a pattern in the light of something humanly significant – and it is therefore not possible strictly to demarcate descriptive and value-bearing statements. 'There are no "naked" facts . . . Accordingly, "appearances" and "values" are not two spheres which admit, as it were, of being spacially removed from one another, and between which there runs a fixed boundary.'[73] All the same, it is for the natural sciences to elucidate the causality in 'appearances' and the disciplines in the humanities to study the

---

72 *Ibid.*, p. 172, and, on Taine, pp. 151–7.
73 *Ibid.*, p. 63.

shaping place of 'values' in human life. The natural sciences cannot subsume the human sciences because there is always place for understanding how the natural sciences, like any other human creation, put into practice an evaluative shaping of the world.

In a manner common in German-language academic culture, Cassirer wrote extensively on intellectual history in support of his philosophy, and in history he traced the process of creating 'form' in the sciences, philosophy and the arts. Historical knowledge, in his texts, is knowledge of human self-formation, and, as such, irreducible to some other knowledge, such as causal knowledge of evolution and inheritance. It is also the route to human self-knowledge. 'History is not knowledge of external facts or events; it is a form of self-knowledge. In order to know myself I cannot endeavor to go beyond myself, to leap, as it were, over my own shadow. I must choose the opposite approach. In history man constantly returns to himself; he attempts to recollect and actualize the whole of his past experience [expressed in symbols].' As a result, 'for understanding and interpreting symbols we have to develop other methods than those of research into causes. The category of meaning is not to be reduced to the category of being.'[74] 'Meaning' is the irreducible subject matter of knowledge of the human subject, and its analysis necessarily diverges from the causal analysis of 'appearances'. Many scholars in the humanities and social sciences have informally identified something like this distinction.

Whether Cassirer, any more than Kant and other German philosophers and philosophical anthropologists, succeeded in establishing the absolute or transcendent rationality of form-giving life was open to question. Ungrounded presuppositions did not disappear. Unresolved tensions, if not contradictions, remained in Cassirer's work, as in Dilthey's or Rickert's, between a Kantian assertion of the necessary, objective 'form' of knowledge creation and its historically relative, particular character. Gadamer therefore subjected the whole relationship between history and philosophy to a searching critique.

74 Cassirer, *Essay on Man*, pp. 191, 195.

He did so by examining hermeneutics as both a scholarly tradition and the site of formal questions in the theory of knowledge.

Gadamer systematically discussed the epistemology and evaluative content of the notion of *Geisteswissenschaft* in *Wahrheit und Methode* (1960). By '*Geisteswissenschaften*', it is important to note, he meant what Anglo-American universities call the humanities. Gadamer's translator chose to render '*Geisteswissenschaften*' by 'human sciences', however, and this had the apparently paradoxical consequence in the text that 'the human sciences are connected to modes of experience that lie outside science: with the experiences of philosophy, of art, and of history itself. These are all modes of experience in which a truth is communicated that cannot be verified by the methodological means proper to science.'[75] Translation into English had the odd sounding effect of making 'science' (in this case, the disciplines of the humanities) lie outside 'science' (in this case, the natural and social sciences).

It is possible to clarify this by saying that Gadamer's concern was the means by which the sciences of the human sphere (known by whatever collective term), the sphere marked by interest in what is true, good and beautiful, can claim truth. In the wake of Heidegger's call for philosophy to re-engage with being, Gadamer criticised the notion that any *method*, whether of a natural or of a social science, could possibly be the means to arrive at the *truth* of human existence. He dismissed the psychological and social sciences' commitment to the authentication of truth by empirical methods as philosophically trivial. Whatever the justification these sciences may have for their huge investment in training and methods, it cannot be the claim that the methods are themselves the road to truth. This road, Gadamer reasserted, has philosophical reflection as its prior condition. Thus his work was a renewed attempt to deal with the late nineteenth-century crisis produced by the explosion of knowledge in the natural sciences, and later by the social sciences, divorced from rational understanding of what making truth claims about being human

75 Gadamer, *Truth and Method*, p. xxii.

actually involves. He turned and re-examined hermeneutics in order to assert the authority of hermeneutics as the *activity of understanding itself* not a *method*.

Gadamer feared that the natural, psychological and social sciences together, by claiming to arrive at truth by an object-ive method, were damaging, if not destroying, the western tradition of seeking truth by the exercise of philosophical, historical and aesthetic reason. He did not criticise the achieve-ments or method of any of the sciences (including the human-ities disciplines). In contrast to many superficial views of the supposed existence of 'two cultures', he firmly believed that there is one method common to all the sciences (again, includ-ing the humanities). 'What is called "method" in modern sci-ence remains the same everywhere and is only displayed in an especially exemplary form in the natural sciences. The human sciences have no method of their own.' This is surely correct – though one might wish to refer to methods, in the plural. What distinguishes the human sciences or humanities is their capacity to reason to a form of knowledge which is different from the form in the natural sciences. Reason is not a method but understanding itself attempting to work from the basis of lived experience and not from facts methodically abstracted from such experience. And it is possible to give an account of understanding, Gadamer argued, by elucidating hermeneutics as 'a theory of the real experience that thinking is'.[76] The his-tory of hermeneutics, which Gadamer proceeded to write, is a history of understanding and not a history of the method of philosophy, history and aesthetics.

Gadamer's overall framework was normative – true know-ledge expresses goals or ends – and he laid out his argument in a discussion of aesthetic judgement (as Kant had done in his third critique). He self-consciously returned to the late eighteenth-century German ideal of *Bildung*, 'the *concept of self-formation, education, or cultivation* . . . which is the atmo-sphere breathed by the human sciences of the nineteenth century, even if they are unable to offer any epistemological justification for it'. And, he went on: 'What makes the human

76 *Ibid.*, pp. xxxvi, 7–8.

sciences into sciences can be understood more easily from the tradition of the concept of Bildung than from the modern idea of scientific method.' Indeed, he thought, if natural or social scientists were consistently to apply their methods, there could be no place for the humanities. 'A logically consistent application of ... method as the only norm for the truth of the human sciences would amount to their self-annihilation.'[77] Without a culture of *Bildung*, he argued, there will be no way to understand or justify the pursuit of truth. By elevating methods, as positivist scientists have done, science will be left with no reason to go after truth in the first place. This is 'self-annihilation'.

The alternative, Gadamer thought, is the process of individual self-formation and cultural self-formation, as exemplified in the creation and encounter with art. What art achieves is the representation of living experience in a manner which transforms all who share it, opening the horizon of reason to an evaluative dimension which has potential as a human universal, an ideal of the human spirit. Like Dilthey and Husserl, in their different ways, Gadamer turned to 'the natural givenness of our existence', the 'units of experience [which] are themselves units of meaning' and not reducible.[78] Art, as also the disciplines of history and philosophy, re-creates these givens, which the Stoics and the Renaissance rhetorical tradition identified as the *sensus communis*, the common weal, into forms which more fully develop their potential. Through hermeneutics, we interpret what these forms of knowledge mean. Carrying this out is the task of the sciences of being human.

Knowledge in the human sciences (i.e. humanities), in Gadamer's account, is, therefore, understanding, and 'understanding must be conceived as a part of the event in which meaning occurs, the event in which the meaning of all statements – those of art and all other kinds of tradition – is formed and actualized'. Whereas it can be said that the natural and social sciences make progress in so far as they increase knowledge of nature or society, the humanities seek to clarify

77 *Ibid.*, pp. 9, 18, 19.
78 *Ibid.*, pp. xxiv, 65.

meaning. As meaning is embedded as 'tradition' – that is, historically formed culture – knowledge of meaning involves a dialogue between the learning which we undertake in seeking *Bildung*, or self-formation, and the learning which others have expressed elsewhere and at other times. This dialogue, or 'conversation' as Gadamer called it, is the substance of hermeneutical enquiry. It is this dialogical activity (to use Bakhtin's term), not subject matter, which marks out the human sciences. This is a key point:

> Obviously, in the human sciences we cannot speak of an object of research in the same sense as in the natural sciences, where research penetrates more and more deeply into nature. Rather, in the human sciences the particular research questions concerning tradition that we are interested in pursuing are motivated in a special way by the present and its interests. The theme and object of research are actually constituted by the motivation of the inquiry. Hence historical research is carried along by the historical movement of life itself and cannot be understood teleologically in terms of the object into which it is inquiring. Such an 'object in itself' clearly does not exist at all.

The constitution of 'traditions', themes which we, in the present, identify as mattering to us, is of central interest in the history of the human sciences. Using them to give form to what we call the past, we make possible reflexive culture about what we understand to be human. Truthful, as opposed to false, reflection requires us to take note of what others were saying. This is the *discipline* of hermeneutics. Where this exists there is 'the fusion of the horizons of understanding, which is what mediates between the text and its interpreter . . . [This] *is actually the achievement of language.*' Hermeneutics as an activity had begun as work on texts. But Gadamer intended that hermeneutics should become the living self-reflection through language that language makes possible. 'We are endeavoring to approach the mystery of language from the conversation that we ourselves are.'[79] The sciences of being human formalise this 'conversation'. 'In the linguistic character

79 *Ibid.*, pp. 164–5, 284–5, 378.

of our access to the world, we are implanted in a process of tradition that marks us as historical in essence.'[80]

These arguments reopened the question of the hermeneutic circle. What can be said about the objectivity of knowledge if our understanding of an interlocutor always presumes some knowledge about what a statement means to her or him? For Gadamer, 'fundamentally, understanding is always a movement in this kind of circle', a condition of all knowledge. As argued earlier in the discussion of the reflexive nature of knowledge, this conclusion is surely correct. The world 'is always a human – i.e., verbally constituted – world that presents itself to us . . . The criterion for the continuing expansion of our own world picture is not given by a "world in itself" that lies beyond all language. Rather, the infinite perfectibility of the human experience of the world means that, whatever language we use, we never succeed in seeing anything but an ever more extended aspect, a "view" of the world.'[81] There is necessarily a place in our knowledge for the way particular uses of language are involved in a 'view' of how particular people, scientists obviously included, have constituted the world. Knowledge which does not include reflection on the sources of its own language is incomplete. This reflection may be formal and abstract or it may be historical and concrete. The latter reflection, on the 'traditions' which shape 'views' of what it is to be human, gives the history of the human sciences its purpose and identity.

Gadamer's work illustrates once again how utterly impossible it is to detach discussion of what the human sciences are from fundamental questions about what is human. Centuries of discussion about the classification of the sciences (in the continental European sense) is talk about human nature and being human. To say what the human sciences *are* is to make a prior commitment, ultimately ontological, about the conditions of knowledge. The positivists hoped to side-step this by establishing an empirical criterion for deciding what knowledge is, and this remains, in practice, the hope of many natural

80 Gadamer, 'Heritage of Hegel', p. 50.
81 Gadamer, *Truth and Method*, pp. 190, 447.

and social scientists who want to get on with the business of knowing the world for practical ends. But if we look at the genealogy of the human sciences we see the re-creation, not the humbling, of fundamental questions.

This chapter has restored to view arguments for the autonomy of the human from the natural sciences. These arguments do not constitute a tradition properly so-called, since they are not connected by continuous institutional or cultural links. I am not interested in myth-making about founding fathers but in the arguments which establish *precedent* for a philosophical position. Authors from Vico to Gadamer, and on into the contemporary human sciences, have argued that human beings understand themselves and the world through using language and, in this process, constitute what they are. The natural sciences, however exemplary their precision and objectivity, do not and cannot disclose knowledge of this.

# 5

# Historical knowledge

## Narrative

Clio has been standing in the wings while actors have declaimed about reflexivity and understanding in the sciences of being human. The drama has been preparing for her move to the front of the stage.

The historical dimension of being human is implicit in reflexive thought; there is no place on which to stand to know 'from the outside'. All knowledge is interpretive – there always are presuppositions, and these are always open to critical examination. If examining the presuppositions of reason appears to lead to self-refuting contradictions, these must be reconsidered as conditions of thought. Historical knowledge of the assumptions and interpretations people have, as a matter of fact, made is fundamental to knowledge. Furthermore, since self-knowledge changes 'the self', and collective knowledge changes communal identity, knowledge of being human centres on the process, or history, which links knowledge and identity. In addition, the failure to provide exclusively empirical criteria for the confirmation (or falsification) of knowledge claims led to a revaluation of the historical creation of knowledge. Finally, history shows that there is substantial precedent for thinking historical knowledge intrinsic to the human sciences. Scholars from Vico to Gadamer have emphasised the historical creation of being human. Though Foucault rejected philosophical anthropology, he too turned to history for his own analysis of the constitution of power and knowledge. Added together, these arguments make it clear why, when the term 'the human sciences' came into common use in the last decades of the twentieth century, it did so as part of a 'historical turn' in the English-language social sciences and humanities.

It is no denigration – quite the contrary – to recognise the roots and continued closeness of historical science to story-telling. The word 'story', in English, is the older word 'history' with a lost syllable, while the German *'Geschichte'*, the Russian *'istoria'* and the French *'histoire'* all denote both history and story. This closeness has much to tell us.

In the late 1970s, the Anglo-American historian Lawrence Stone noted a 'revival of narrative' in historical writing.[1] This was manifested in a turning away from empirical social science as a model for the kind of work which historians should be doing. Whereas, a decade earlier, there had been enthusiasm, especially among social historians, for assembling quantitative data as the basis for causal generalisation, historians began to reassert their interest in narrative about particular events. Narrative was in fact a form many historians had never surrendered, but they began to give positive reasons for writing narrative rather than to respond defensively to a supposed failure on their part to explain events by causal generalisations.

In practice, the distinction between narrative and causal explanation was rarely clear cut in historical writing, just as the distinction between understanding and explanation was not clear cut in social science. Analysis showed that narrative and explanation are not mutually exclusive; indeed, they frequently run together, though authors obviously also stress one rather than the other to suit their purposes. The underlying reason is that any explanation, since all explanations are interpretive, makes sense only by virtue of its place in a narrative (whether explicit or implicit).[2] Some social scientists and social and economic historians, however, thought the revived interest in narrative retrograde, since it appeared to consign history to a position where it is not a real, explanatory science. They had a point, in a way, since many historians were much more at home with the 'art' of creating a story about particular people or events than with the 'sciences' of causal explanation.

---

1 Stone, 'Revival of narrative'.
2 See Ricoeur, 'The narrative function'.

But the critics were wrong in a deeper sense: narrative knowledge is scientific knowledge, scientific knowledge of a form appropriate for understanding the kind of being that is human. It was the authority of the standard model of scientific knowledge based on the physical sciences which misled historians into thinking that only a certain kind of causal explanation could establish the scientific credentials of their field.

Philosophers who had opposed positivism and behaviourism had established the legitimacy of narrative in history by giving prominence to the stories, or histories, which people tell in making sense of action. Hampshire wrote that talk itself implicates history: 'In order to communicate, we need to have the means of referring to elements in reality which have a history, that tells how they came to be here, standing in this relation to this speaker as objects of reference.'[3] Similarly, MacIntyre observed: 'Narrative history of a certain kind turns out to be the basic and essential genre for the characterisation of human actions.'[4] Taylor took it to be a 'basic condition of making sense of ourselves, that we grasp our lives in a *narrative'*.[5] Earlier, Heidegger had reflected on the temporality of the human condition, implying that historical narrative has irreducible importance.[6] As a consequence, phenomenologically oriented studies argued, that 'narrative form is not a dress which covers something else but the structure inherent in human experience and action'.[7] Behind narrative as discourse, phenomenologists discerned narrative as representation of being: 'In other words, *the form of life to which narrative discourse belongs is our historical condition itself.'*[8]

Anglo-American historians rediscovered in the narrative form the mode of writing history concerned with the meaning, as opposed to sequence, of events, and they understood meaning

---

3 Hampshire, *Thought and Action*, pp. 17–18.
4 MacIntyre, *After Virtue*, p. 194.
5 Taylor, *Sources of the Self*, p. 47.
6 Heidegger, *Being and Time*, pp. 38–45.
7 Carr, *Time, Narrative, and History*, p. 65.
8 Ricoeur, 'The narrative function', p. 288. Also Ricoeur, *Time and Narrative*.

as arising out of a linguistic act.[9] According to analytic critics
of positivism and philosophers in the hermeneutic tradition,
it cannot be said that a statement has meaning because it
describes the world. Rather, meaning derives from the expressive, and historically specific, act of describing the world. This
act, as historians have generally recognised, involves selection
among the possible objects of knowledge. No history of an
event attempts to recount all the causes of what happened;
this is in principle impossible, since the whole history of the
world is in some small measure a cause. It was Collingwood
who made the profound point that 'everything in the world is
potential evidence for any subject whatever'.[10] Selection is
therefore a constitutive part of creating meaning.

Historians decide what is significant, and they do this by
locating an event or action, and its causes, in a narrative or
story. Which story the historian chooses – or, as it may be,
takes for granted – depends on the historian's purposes. 'To
ask for the significance of an event, in the *historical* sense of
the term, is to ask a question which can be answered only in
the context of a *story*. The identical event will have a different
significance in accordance with the story in which it is located
or, in other words, in accordance with what different sets of
*later* events it may be connected.'[11] The judgements about significance which historians bring to their work structure how
they tell and plot stories. 'The historian arranges the events in
the chronicle into a hierarchy of significance by assigning
events different functions as story elements in such a way as
to disclose the formal coherence of a whole set of events considered as a comprehensible process with a discernible beginning, middle, and end.'[12] The narrative form therefore builds
a kind of teleology into the events recounted, since it makes
those events fulfil a purpose and reach a conclusion, and that
purpose and conclusion is what the historian has picked out
at the beginning. As a result, writers reintroduce teleology in

9 Such as Bouwsma, 'From history of ideas to history of meaning'.
10 Collingwood, *Idea of History*, p. 280.
11 Danto, *Analytical Philosophy of History*, p. 11.
12 White, *Metahistory*, p. 7.

their language, even where, as in stories about evolution, it is the point of the story to rule teleology out of the account of what happened. 'The enterprises of scientist and of writer both act out the paradox that narrative implies teleology even when its argument denies it.'[13]

In one existentially crucial sense, it should be noted, the plots of stories and the plots of histories differ, since there is always a sense in writing history that the end has not yet been reached.[14] Moreover, according to the reflexive argument, what we now emplot as the past has a place in determining what the future will be.

Writers of literary narrative and everyday story-tellers, of course, had never forgotten that they were in the business of judging significance. What was not always so clear was that the narrative form affects the content of knowledge and not just the style of presentation. Different narrative modes build presuppositions, judgements and norms into what is said, in scientific writing as in story-telling. Narratives, analysts showed, are theories, though not the kind of general law theories which social scientists, and historians who followed their example, thought they were searching for. They are theories with local scope. 'Narratives may be regarded as kinds of theories, capable of support, and introducing, by grouping them together in certain ways, a kind of order and structure into events. A narrative, so considered, is nevertheless localized as to space and time, it forms an answer to an historical question, and is accordingly to be distinguished from a general theory.'[15]

A vivid way to grasp the importance of narrative is to observe what discourse is like when narrative is absent, or at least when its commonly accepted forms are absent. The speech or writing of people with schizophrenia, for instance, exhibits 'the lack of a cohesive theme or narrative line' and for this reason appears 'crazy'. Without a narrative determination of significance, we cannot enter another person's world. The

---

13 Beer, 'Problems of description', p. 50.
14 See Steedman, *Dust*, pp. 142–56.
15 Danto, *Analytical Philosophy of History*, p. 137.

language of schizophrenia 'may also sound telegraphic, as if a great deal of meaning were being condensed into words or phrases that remain obscure because the speaker does not provide the background information and sense of context the listener needs to understand'.[16] This calls to mind modernist aesthetics, which also sought to escape from conventional narrative forms. Modernist writers, when they went beyond narrative, created writing which the audience had to be familiarised with, had to learn how to place in a narrative about what the artist was up to, for it to appear meaningful. Wanting to escape conventional bounds of meaning, artists have searched repeatedly for ways to break audiences of the habit of placing art within a narrative. The choreographer Merce Cunningham, for example, used randomising techniques to determine dance sequences.

Most historians have a marked preference for empirical discourse; this is built into the ethos of the discipline through the relationship which historians have with what they call the primary sources and the archive. Narrative, as a manner of writing which gives a voice to the particularities of primary sources, has thrived in historical scholarship. A commitment to empirical narrative, focused on particulars, was central to the establishment of a professional field, in German scholarship in the first half of the nineteenth century and a little later elsewhere. Training in writing on the basis of primary sources became the core of practice. Narrative was the means to turn this writing into coherent, intelligible and, not least, interesting knowledge. 'Historical understanding is the exercise of the capacity to follow a story, where the story is known to be based on evidence.'[17] Thus encouraged, historians, most of whom wanted to get on with historical research rather than to reflect on the nature of historical knowledge, wrote as if the narrative voice simply re-creates the content of the sources or, even, re-creates what ordinary language calls 'the past'. However, as Collingwood (drawing on the work of Michael Oakeshott) rightly concluded: 'The historical past is the world

16 Sass, *Madness and Modernism*, pp. 156, 177.
17 Gallie, *Philosophy and the Historical Understanding*, p. 105.

of ideas which the present evidence creates in the present. In historical inference we do not move from our present world to the past world; the movement in experience is always a movement within a present world of ideas.'[18]

Indeed, though historians rhetorically make the primary sources the authoritative voice in narrative, they are not blind to the fact that what they write originates with their own questions and purposes, with selection. What ordinary language calls the past is the result of the kind of dialogue made familiar in studies of hermeneutics. 'On the one hand, the real is the result of analysis, while on the other, it is its postulate. Neither of these two forms of reality can be eliminated or reduced to the other. Historical science takes hold precisely in their relation to one another, and its proper objective is developing this relation into a discourse.'[19] In the case of the history of the human sciences, 'the real' is being human, and it is both object and agent of the reflexive act. Historical writing in the human sciences takes up the relation between 'the two forms of reality' and thereby attempts to make reflexive action clear to itself.

In the last third of the twentieth century, a new interest in what Collingwood called 'movement within a present world of ideas' encouraged philosophers of history, like Hayden White, to turn to literary theory for insight, there to discover, in the analysis of rhetoric, poetry and fiction, a wealth of resources for the analysis of non-fiction. Like the literary writer, the historical writer begins with a shaping act, with *poiesis* or production.[20] Some historians responded with anger because they read such work as saying that history is a form of literature, impugning, these critics believed, the distinction between fact and fiction which gives disciplinary history both its cognitive and its moral *raison d'être*. But what White claimed was that history is knowledge, and, as with other kinds of knowledge, the historian must prefigure subject and meaning if the object of knowledge is to be given a comprehensible representation.

18 Collingwood, *Idea of History*, p. 154.
19 Certeau, 'Making history', p. 35.
20 White, *Metahistory*, pp. 30–1. Also White, *The Content of the Form*.

(I have already referred to Heidegger's expression of this argument.) He also showed in detail how the poetic forms of literature, such as tragedy and comedy, provide historians with the forms of their own writing when they create intelligible stories. There is no 'real past' to be uncovered; there is, however, massive primary resource material for the historian to render in literary form.

The same argument extends to natural science writing. At first glance, there is an unequivocal contrast between the research paper in the natural or social sciences and the historical story which, if an attempt were made to put it a research paper, would be rejected as anecdote. Indeed, the language of the scientific paper may be almost entirely mathematical. But is this contrast absolute? The examination of scientific writing suggests that it is not, since scientific writing, like literature, has its rhetoric, narrative form, plot and voice.[21] The difference is that formal scientific papers conform to a set of rigid conventions which police objective argument. Each discipline has, over time, built up elaborate rules to control the quality of what researchers say, and it is possible to analyse scientific writing and show how writers follow these rules in order to signal the presence of authoritative knowledge. 'The intelligibility, significance, and justification of scientific knowledge stem from their already belonging to continually reconstructed narrative contexts supplied by the ongoing social practices of scientific research.' The fact that writing in science, or in history, aims at truth (when it does), does not eliminate either the poetic features of language or the rule-bound tradition to which the writing is a contribution. Science writing *disciplines* these features through what researchers know as the evidence. Further, disciplinary rules usually detach writing from the history of how this kind of writing came to be current and comprehensible and from the wider purposes which it serves. The scientific paper is therefore intelligible only to specialists

21 Bazerman, *Shaping Written Knowledge*; Bazerman, 'Codifying the social scientific style'; Myers, *Writing Biology* – for sociobiology texts; Nelson, Megill and McCloskey (eds), *Rhetoric of the Human Sciences*.

who have considerable tacit knowledge. Scientific papers 'do not *tell* stories about how the current research situation came about and where it seems to be heading. They are instead more complicated speech acts, which can only be understood in the *context* of a story [which the paper does not spell out].'[22]

Different kinds of writing share in common the narrative, historical constitution of the world; the history or story is not 'out there' and described by narrative but 'in there' in the narrative. This suggests that historical writing is not just one genre among others which we can adopt or discard at will. History is built into the form of knowledge, the narrative, in terms of which people state meaning, assign identity and express purpose. Narrative, in Louis Mink's words, is an 'irreducible form of understanding'.[23] The story, or history, gives language the capacity to make sense. The study of history, therefore, is not at root academic, but academic work disciplines a fundamental dimension of human life. 'The historical sciences only advance and broaden the thought already implicit in the experience of life.'[24] Collingwood wrote simply that history is not 'a mode of experience, but an integral part of experience itself'.[25]

### Narrative purposes

When scientists, including historians, describe and explain events, they pursue ideals, like objectivity, by acting according to standards built up and exposed to criticism over many generations. Some standards are so taken for granted in the way of life of being a scientist, and so embedded in institutionalised structures rather than explicit rules, that they become visible only when specifically broken, examined or criticised. Sociologists of science have investigated how such standards operate. One result is that scientists sometimes feel that the purposes of science itself are called into question; but

22 Rouse, 'Narrative reconstruction of science', pp. 181, 188.
23 Mink, 'Narrative form', p. 132.
24 Gadamer, *Truth and Method*, p. 222.
25 Collingwood, *Idea of History*, p. 158.

they are certainly mistaken, since sociologists are scientists too and also aim objectively to understand the world.

The conditions of objectivity in historical knowledge are the same as in any hermeneutic activity.

> [The historian] cannot simply jump over the open horizon of his own life activity and just suspend the context of tradition in which his own subjectivity has been formed in order to sub-merge himself in a subhistorical stream of life that allows the pleasurable identification of everyone with everyone else. Never-theless, hermeneutic understanding can arrive at objectivity to the extent that the understanding subject learns, through the communicative appropriation of alien objectivations, to com-prehend itself in its own self-formative process.[26]

The ideal of objectivity involves a search for knowledge of self as well as knowledge of what is other. Without knowledge of the former, it is not possible to understand the interpretive process through which scientists gain knowledge of the latter. In normal academic life, however, this self-knowledge takes the form of disciplinary conventions, which tend to become shared hidden assumptions rather than personal insights.

In history, these assumptions about the dependence of what historians say, or should say, on the basis of the appropriate primary sources lead to a self-conscious distinction between fiction and non-fiction. What distinguishes them is a claim about truth; but this is no absolute distinction, since people commonly tell true stories and many historians face criticism for not getting at the truth. I do not, however, want or need to get into the arbitration of the distinction between fiction and non-fiction. All that is needed is to establish that scientists, including historians, adopt forms of writing suitable to the purposes they have in view. Both the forms of writing con-sidered appropriate for these purposes, like truth-telling, and the purposes themselves are embedded in the conventions of the community to which scientists belong.

This is clearly visible in the old argument between his-torians of science writing as natural or social scientists and historians of science writing as members of the history

---

26 Habermas, *Knowledge and Human Interests*, p. 181.

profession. The interests of natural or social scientists in tracing the sources of modern knowledge and in supporting the ideals of natural or social science dominated earlier writing. Around 1970, a new generation of writers, more interested in historical scholarship, pushed for much more contextual plots, that is, plots built up around the meaning that knowledge had for actors in their own time. Historians promoted contextual explanation of the meaning of texts, and they criticised what they called 'presentism', characteristic of a genre in which current scientific interests dominate the plot. The new historians of science adopted the empirical conventions of the mainstream history discipline and wrote as if something called the past determines the content of historical knowledge. The resulting conflict between different narrative forms reflected the purposes of different writers. Historians of science and scientists who wrote history retained an interest in different narratives, and each narrative contained a potential critique of the other. But if the purpose was to write *historical* narrative, then authors adopted the standards of the history discipline, took account of the full range of evidence and embedded meaning and explanation in context. This still left them with a great range of possible narrative forms.

Subsequently, academic historians of science became self-conscious about the philosophical limits of anti-presentist argument, by now almost a cliché in their discipline.[27] The result was the rediscovery of the hermeneutic principle that the point of writing, the 'view', necessarily informs what is written.[28] This overlapped with political critique: all vantage points of writing occupy one place in the social and political world rather than another, albeit that the vantage point may be that of a well-established science. Writing involves selecting facts on the basis of judgement about significance: the idea of a history of everything is a standing joke. It is also suspect. An over-exclusive emphasis on facts produces writing

---

27 Blondiaux and Richard, 'A quoi sert l'histoire des sciences de l'homme?'; Kelley, 'Problem of knowledge', p. 25.
28 See Markus, 'Why is there no hermeneutics of natural sciences?', p. 7.

which 'resembles statistics, which are such excellent means of propaganda because they let the "facts" speak and hence simulate an objectivity that in reality depends on the legitimacy of the questions asked'.[29]

We cannot use causal relevance, by itself, as a criterion of selection; what is causally relevant is given in a pre-existent plot.[30] Only if we have a philosophy of history in the eighteenth-century sense, faith that human existence works itself out according to a transcendent plot, is it possible to say that there is an absolute criterion of relevance about which facts matter. Only then could we hope to give a historical description which, in Humboldt's words, 'puts the isolated fragment into its proper perspective' and permits 'the representation of the form of both the universal and the individual existence of natural objects'.[31] Some scholars in the modern Islamic world, for example, advance such a philosophy of history; so do western science writers who represent the advance of science as the triumph of an exclusive truth. By contrast, the literature on historiography – the analysis of what historians have written – is one long argument about *what* facts matter for *which* plot.

Comparing a personal life story and a history is very revealing. The narrative of a person's life from birth to death, with space for the plots of romance, comedy, tragedy or satire in between, has a profound hold on the imagination. This storytellers have long known and the popularity of biography and autobiography has long confirmed. Dilthey was surely right in believing that the question of how to write autobiography goes to the heart of human self-understanding. We confidently assert that the ability to tell a story about a life, whether our own or another person's, gives meaning to that life. MacIntyre has especially stressed this: 'It is because we all live out narratives in our lives and because we understand our own lives in terms of the narratives that we live out that the form of narrative is appropriate for understanding the actions of others.'[32]

29 Gadamer, *Truth and Method*, p. 301.
30 See Mink, 'Narrative form', pp. 133–5.
31 Humboldt, 'On the historian's task', pp. 58, 59.
32 MacIntyre, *After Virtue*, p. 197.

It may be that the plot of the individual life is the one plot of depth and power left to people for whom the great plots of religion, and the re-expression of those plots in schemes for the redemption of the world through social progress, have been destroyed. A claim about who we are and where we are going makes sense, in a disillusioned age, only in terms of a plotted individual life.

Psychoanalysis and related forms of psychotherapy vividly illustrate the importance of the story as a source of meaning. Classic analysis was a complex linguistic game, a specially crafted social encounter designed to produce certain kinds of stories about a person's life. Acting out the story in the symbolic terms of the transference became the means to reshape 'the self'. Freud famously conceded that 'the case histories I write read like novellas, and . . . they, so to speak, lack the serious stamp of scientific method'.[33] Critics seized on this admission to claim that even Freud knew he was no scientist. All the same, it can be thought, looking at the impact of his work, that it was precisely this way of writing which was his strength and not his weakness. The power of these stories was not their empirical veracity, nor that they followed a formal method, but that they enabled patients, therapists and readers to find meaning, and hence direction, in modern lives. There was a stark contrast between the 'talking cure' which Josef Breuer and Freud offered to hysterical women and psychiatry's dismissal of ill people's speech and expression as meaningless. By its willingness to construct stories where none had been permitted before, especially about intimate desire, psychoanalysis offered new possibilities for identity.

The obvious way to make clear the purposes of story-telling and to bring out the implications stories have for creating one way of life rather than another, is to tell a story.

### Telling a likely story

In 1845, Martha Brixey, an unmarried young woman, took a knife and severed the head of a small child in the family in

33 In Breuer and Freud, *Studies on Hysteria*, p. 160.

London where she worked as a servant. She immediately told her employer what she had done and asked whether she would be hanged. Tried at the Central Criminal Court for murder, she was acquitted on the grounds of insanity under the McNaughtan Rules, introduced in England in 1843, and she spent thirteen years in the criminal lunatic wing at Bethlem hospital. Then she was released as 'a young woman of most interesting appearance and amiable expression of countenance with great propriety of manner and a pattern of retiring, modest behaviour. There is no appearance of anything approaching insanity or delusion of any kind.'[34]

This is a true story, or history, of a defendant using the insanity defence to answer an accusation of murder. People tell very different stories about why people who have killed or committed other crimes do what they do, and people told different stories about this event; and different historians might now also retell the story differently.

Both codified and common law criminal systems make provisions for a defendant to argue that he or she is not criminally liable because the offence was not the act of a sane person. The precise wording of the defence, and the circumstances in which it is right for the court to uphold it, have been the source of huge debate. As the size of debate signifies, what looks to be a matter of factual decision – was the defendant insane when he or she committed the crime? – involves complex conceptual, legal, ethical, medical and political issues.

Two stories featured at Brixey's trial and in Victorian commentary on the case. In the first, she had consciously planned a terrible act, which shows the blackness of the human heart, taking a sharp knife from the kitchen. Such an act in a respectable home was threatening and shocking and people had to be held responsible. There was a danger which must be eliminated. By contrast, some doctors argued that her menstruation was suppressed, she was overcome by a fixed idea and, while committing the crime, she could not have known

34 Quoted from Bethlem records in Smith, 'Mad or bad?', p. 14.

what she was doing. What happened was therefore an exceptional breakdown of a physical system not an attack on hearth and home. Mixed in with this medical argument was the ordinary person's thought, which perhaps played a role at the trial (though it was not a legal argument), that someone who does such a literally extraordinary thing cannot be sane. What the defendant herself thought is hard to know, and no one was much concerned with it, but she did make the pertinent comment that she had hoped to make an angel of the child. This was a sentiment present at times in other somewhat similar cases.

The story which attributed guilt pictured the defendant as an agent. In this story, she had reasons, however evil, perverted and abnormal, and she executed a plan and remained fully conscious and aware of what she had done and of its illegal nature. She personified the evil and anarchy lurking in human nature against which the law was, or should be, a bastion. In terms of causal attribution, this narrative placed weight on the person as moral agent. If she had felt such desires, she could have acted to create circumstances where they could not have had such consequences. The medical narrative, however, significantly speaking of the defendant in the passive voice, spoke of the weakness of a woman's body and the disturbance this causes in the brain. The story followed a standard pattern of causal attribution. A disordered body caused what happened, it acted on the defendant and she was merely the vehicle for what was done to the victim.

Both the retributive and medical narratives now look rather feeble, though the same kind of questionable causal attributions appear in what we might think of as more socially sensitive modern stories. We might, for example, feel sympathy with the position of an unmarried servant, and perhaps a chaste young woman, who would lose her position on marriage and whose idealism contrasted the innocence of children with the often sordid prospects of adults. More plausibly to my mind (though with no confirmation in this case), this may have been an example of a not uncommon 'nostalgia' among young servants. Torn from home into alien and emotionally unresponsive environments in order to earn a living, young people sometimes reacted with extreme distress.[35] All the same, to attribute causes to social failings is, equally with the medical story, to make the offender an object not a subject of what happened.

The court had to reach a decision, and one story rather than another served to legitimate and make sense of that decision. The court carried out in formal and dramatic circumstances what goes on all the time in everyday life when decisions are taken about causal attribution and responsibility for car accidents, being lazy, unemployment, a child's misbehaviour or whatever. There is always the potential for different stories, and we must choose. 'We cannot refer to events as such, but only to events *under a description*; so there can be more than one description of the same event.'[36] Habit and convention, however, usually bury the choice in rules about which empirical evidence, in what quantity or quality, legitimates which story. The same forms of narrative attribution involving causes and intentions are available to describe all people, whether virtuous, mad, wise, children or criminal. No amount of empirical information in itself can constitute grounds for a decision about which explanation to use, though certain kinds of information indicate that it is a social occasion to use one explanation rather than another. In normal circumstances, members of the same society share rules about which explanatory narrative to deploy on which occasions. The controversial court case exposes the limits of convention. As MacIntyre observed: 'I can only answer the question "What am I to do?" if I can answer the prior question "Of what story or stories do I find myself a part?"'[37] In which story did Brixey have a part?

The historian is in a similar position to the ordinary person or the criminal court. The historian, too, selects which story

---

35 See Starobinski, *Action and Reaction*, p. 205, citing the authority of Karl Jaspers and the model case of Johanna Spyri's *Heidi* (1880).

36 Mink, 'Narrative form', pp. 145–6.

37 MacIntyre, *After Virtue*, p. 201. Also Barham, *Schizophrenia and Human Value*, pp. 47–99.

to tell, and does so as a person living in a particular community – as an academic historian, as it may be – and as a member of other social groups, all with histories and tacit values of their own. When I tell Brixey's history, I juxtapose different stories that people told, and in this I display two purposes: the historian's intent to give a coherent picture of evidence in the primary sources, and the historiographer's intent to analyse what is going on when the historian constructs narrative. Many historians, like social scientists, would add that the final goal is causal explanation. For a single, singular event like Brixey's crime this is particularly difficult, since neither medical psychology nor moral judgement seems a particularly persuasive source of causal insight in such cases. The historian has to decide which causes, from a possible infinite range, matter; that is, the historian forms causal explanation into a narrative. Talk about the relations between stories, including the historian's story about other stories, makes a reflexive narrative. Literary writers have long explored this and the way differences in style, voice and form of telling affect what is said. The French writer Raymond Queneau told the same trivial story in ninety-nine different versions in order to illustrate the riches of style![38]

Jurists have long tried to make clear what separates a legally culpable act and a deed not entailing liability. Interestingly enough, the legal capacity for action has not matched exactly with human action, and thus the legal distinction between act and deed does not correspond in all cases with the 1960s philosophical separation of reasons (resulting in acts) and physical causes (resulting in deeds). For example, the earlier English common law of deodands made it possible for a cart to be held liable for an accident and hence forfeited, or an animal, which had caused injury, to be beaten or killed. Xerxes, son of Darius, ordered the sea to be scourged with rods for having engulfed his fleet. Long before modern analytic philosophy, jurisprudence, religion and poetry had established narrative patterns mapping agency in the world. This was the repertoire which commentators on Brixey's case drew on.

38 Queneau, *Exercises in Style.*

Stories about crime continuously re-create the world in which people can be held responsible for what they do, in which wickedness and goodness are present and in which political and social intervention makes sense. But these stories also tell about the limits to what people can do.

All this reinforces the main point which I am driving at: any description of what it is to be human has content and meaning by virtue of its place in a narrative. There is no absolutely objective rightness about any particular narrative. The judgement of rightness is a judgement of what is right within the way of life which the narrative helps to make sense of. Each narrative deploys a conceptual scheme, and 'the choice of a conceptual scheme *necessarily* reflects value judgments, and the choice of a conceptual scheme is what *cognitive* rationality is all about'.[39] People select between alternative narrative possibilities as a result of their place within a historical community and tradition and the way, in the freedom of their own minds, they reflect on that tradition. It is the same whether the narrative is part of natural science, history or everyday conversation about who and what people are.

We also come back to a point made earlier in relation to the difference between explaining by physical causes and understanding by reference to social rules. No amount of knowledge of physical laws, the brain, evolutionary history or the genes is sufficient for knowledge of what it is to be human. For the same reasons, no knowledge of economic conditions, childhood deprivation or poor education can fully explain what a person does, however much such realities are conditions for the action. It is always also necessary to explain action in terms of the way human beings represent the world to themselves, in language and in their symbolic life; that is, how they define, imagine and picture what they do.

What authority particular narratives have, whether they are expressed in the language of natural science, social science, religion or what people label common sense, depends on the status and power of each way of framing events in particular contexts and for particular purposes. Authority is a historical

---

39 Putnam, *Reason, Truth and History*, p. 212.

and political matter. It is not at all the case that empirical narratives always have authority, however much the rhetoric of many areas of life gives priority to such narratives, and however much scientists think such narratives should have authority. This is a contingent feature of life. In addition, we cannot rationally decide the place of empirically based narratives, whether in natural science or social science, without first committing ourselves to one way of life rather than another; that is, locating ourselves within one rather than another historical narrative. This is what Brixey's story illustrates. Those who wanted to find her guilty gave priority to the story of a Christian community's fight against evil; those who wanted to find her insane gave priority to a story about empirical knowledge bringing order to human affairs in a modern state. We might want to call the first story traditional and the second modernising. In contrast to both positions, I wrote a didactic history, giving priority to the analysis of story-telling.

The history comes first. MacIntyre put the point emphatically: 'Natural science can be a rational form of enquiry if and only if the writing of a true dramatic narrative – that is, of history understood in a particular way – can be a rational activity. Scientific reason turns out to be subordinate to, and intelligible only in terms of, historical reason.'[40]

### 'What is man?'

Kant's question 'What is man?' has turned into the question 'In what story, told with what purpose, does being human have a part?' It does not help to say that what we want is the truth not a story. The criteria which people, including scientists, use to determine what is truth and what is real are already given embedded in stories. The analysis of narrative has led to the conclusion that historical knowledge is both intrinsic to self-reflection and to understanding what it is to be rational.

This conclusion would seem to come close to Marx's claim (quoted earlier) that 'history is the true natural history of

40 MacIntyre, 'Epistemological crises', p. 464.

man', or even to Ortega's assertion that 'Man, in a word, has no nature; what he has is . . . history'.[41] Marx voiced his belief that it is the human material environment, the product of human activity not nature, in the sense of something existing independently of human activity, which shapes what people are. Later, when he read Darwin, he enthusiastically accepted human evolution, but he continued to think of change in the human world as change in the mode of production – historical, not biological, change. Ortega, by contrast an idealist, returned to a position like Hegel's and proposed that 'man . . . has no nature' because what he has, or rather is, is historically unfolding spirit. Collingwood's formulation – 'the so-called science of human nature or of the human mind resolves itself into history' – though also idealist, was more careful.[42] It did not dismiss the value of biology as Ortega appeared to do, but it did propose to treat biological knowledge as itself an expression of historical life.

These different phrasings re-express the position found in Vico: it is the nature of humanity to make itself over historical time. Thus expressed, the position is in stark opposition to both biological and religious formulas, which attribute the nature of humanity to events, either in evolution or in creation, occurring before history. But all these phrases tend to reinforce the misleading supposition that scientific understanding seeks to delineate the essential characteristics of being human. To the contrary: the more properly scientific project is to understand *concrete* reflexive processes and the manner in which *particular* human groups have created, and may again re-create, identity, meaning and value in their ways of life by telling stories about that life. Talk about the meaning of life needs history not universals. As White argued: 'Meaning is not a constant, not a universal, does not remain the same, but itself changes, in a word, has a history. And this is why our time requires not only a historical understanding of our present

41 Marx, 'Economic and philosophical manuscripts', p. 391; Ortega, 'History as a system', p. 313 (originally in italics throughout; ellipses in the original).
42 Collingwood, *Idea of History*, p. 220.

and recent past, but a new conception of what historical understanding itself consists of.'[43]

Vico proposed to make his new science 'a *history of human ideas*, which forms the basis for constructing a metaphysics of the human mind . . . Hence, the queen of the sciences, metaphysics, began when the first men began to think in human fashion, not when philosophers began to reflect on human ideas.'[44] This argument makes the history 'of human ideas' the science of the human mind as self-creating agent, not the history of discovery of human nature as an entity given in a somehow independently knowable nature.

Here I should comment on two objections. Critics of views such as these sometimes label, and dismiss, them as 'historicist', on the grounds that they conflate the historical understanding of being human with truth claims about what it is to be human. The German scholars who introduced the term in the early twentieth century used it to denote 'the belief that an adequate understanding of the nature of any phenomenon and an adequate assessment of its value are to be gained through considering it in terms of the place it occupied and the role which it played within a process of development'.[45] There are three points associated with this which require to be distinguished. The first is the claim that we think building on the foundations, in culture and language, of what people have thought and written before. 'In the linguistic character of our access to the world, we are implanted in a process of tradition that marks us as historical in essence.'[46] This I accept. The second is the belief that the way of thinking at a particular time is part of a pattern or process of development and, therefore, to understand any particular way of thought is to understand and to value its place in that pattern or process. This position, which amounts to belief in a philosophy of history, was standard among nineteenth-century German scholars,

---

43 White, 'Commentary', p. 133.
44 Vico, *New Science*, p. 128 (§ 347).
45 Mandelbaum, *History, Man, & Reason*, p. 42 (italics removed). Also Schnädelbach, *Philosophy in Germany*, pp. 34–40.
46 Gadamer, 'Heritage of Hegel', p. 50.

like Hegel in philosophy and Droysen (quoted earlier) in history. I have been arguing, however, for a contextual theory of meaning and not for a philosophy of history. The last point, implicit in some references to 'historicism', is belief that the term describes a philosophy of historical determinism: the past has us in an iron grip. Again, nothing I have said supports this; and, in the epilogue, I will argue to the contrary and support a notion of freedom.

The defence of the historicity of knowledge – 'that all distinctions, all rational inquiry and claims, obtain within the bounds of history; they must be subject *to* its flux and cannot exit *from* it' – is properly a philosophical task.[47] Like Margolis, whose characterisation I have just cited, I have turned to Gadamer for authority. What Gadamer, following Heidegger, clearly understood, to repeat, is that we can read a text, or understand the world, only in terms of a 'fore-projection' onto the text or the world. This 'fore-projection' is the 'prejudice' that we bring to bear through our historical position as interpreters with a certain kind of linguistic usage.[48] If this is historicist argument, so be it. But we should note that Gadamer also insisted on historical and philosophical reflection on what that 'fore-projection' is. This rational and, I think, ethical principle opens historicist knowledge to change.[49]

The second objection is closely related: language represents an objectively real world. (In the introduction, I gave reasons for putting the question of realism to one side, at least for the time being.) This is the belief of 'common sense' and the belief taken for granted in many areas of the sciences, where progress, it would seem, demonstrates its validity. It is the philosophy underlying claims by biologists to have given answers, in knowledge of evolution, genetics and the brain, to the question 'What is man?' The response to this, from the vantage point developed here, is to point out that biologists, too, write narratives and that these narratives make sense and command authority because they occupy a place within

47 Margolis, *Flux of History*, p. 99.
48 Gadamer, *Truth and Method*, pp. 266–71.
49 Gadamer, 'Heritage of Hegel', pp. 50–1.

yet other narratives about the nature of objective reason and why objective reason is of value in the first place. Whatever the biological facts may be, the facts in themselves cannot render meaningless the enquiry into the conditions of knowledge and the theory of meaning. In contrast to the realist theory of meaning, we may think of meaning as continuously achieved and re-created in the historical particulars of the collective life we actually have, the life of the institutions of biological science included. As the philosopher of history W. B. Gallie wrote, 'it is a mistake to imagine that any human activity can be understood in a functional yet entirely non-historical manner, i.e. in terms of the specific way in which it fulfils its allotted task, without any reference to how and why *that particular* way came to be preferred or accepted'.[50]

Whatever further discussion of these philosophical questions will achieve, there is no doubting the difference between writing by natural and social scientists who start from a given reality (however minutely they investigate its precise nature) and historians of science who do not take the same given reality for granted. A significant illustration of this is that historians are more likely to argue for conceptual discontinuities in knowledge across the centuries. The historian, to adapt an argument of Janet Coleman, is radically modern in a way that the natural scientist is not, because the former and not the latter considers reason and its fundamental categories as creations of human history. In the romantic positing of 'man' as the creative source of knowledge, reason surrendered the possibility, which informed medieval and ancient culture, of a trans-human, transcendental truth.[51] As Dilthey wrote, in romantic mode: 'When it seems that the soul is about to succeed in perceiving the subject of the course of nature itself, without its masks and its veil, then it discovers in that subject – itself. In fact, this is the final word of all metaphysics.'[52] For Dilthey, the act of knowing discovers itself. But natural scientists, and social scientists who have followed their lead,

50 Gallie, *Philosophy and the Historical Understanding*, p. 222.
51 Coleman, *Ancient and Medieval Memories*, pp. 593–7.
52 Dilthey, *Introduction to the Human Sciences*, p. 321.

however paradoxical it may sound, do not belong (in this sense) to the modern age. They write about the world, rather as a medieval scholar would have done, as if statements transcending the historically embedded construction of truth were possible. They take basic categories, like 'psychology', 'society', 'mind' and 'individual', to lie outside time and to constitute natural kinds. With this presumed, scientists then write history with a teleological narrative, the history of the progressive movement of knowledge towards the truth about the supposedly real entities denoted by such words. But around them there is a modern, or perhaps postmodern, world, to which science as technology has decisively contributed, in which notions of transcendent truth are without moorings and in which we cannot define entities independently of the descriptions, always open to revision, which, as a matter of fact, we give them.

Narrative is a mode of writing which gives form to both everyday and scholarly historical knowledge. Narrative is the verbal expression of the author's own existence as a historical subject, and narrative builds this position into knowledge. 'Our going back to "the past" does not first get its start from the acquisition, sifting, and securing of . . . [historical records]; these activities presuppose . . . the historicality of the historian's existence.'[53] The search for the conditions of knowledge returns to the reflexive action of the present.

53 Heidegger, *Being and Time*, p. 446.

# 6

# Values and knowledge

## Facts and values

Explanation appears in something of a dilemma: we can comprehend events by placing them in a story, which makes sense,
of human invention; or we can explain by the laws or regularities of things, which exist independently of human invention
and have no meaning. At first glance, since it is precisely the
business of science to be objective, the second option appears
right, while the first appears guilty of anthropomorphism, projecting what is distinctive of being human where it does not
belong. Yet, it is narrative, most certainly a human invention,
which makes statements about the world, including statements
of natural laws and regularities, meaningful or comprehensible.
And narrative introduces a plot, or purpose, and judgements
of significance into nature where, objectively, they appear
not to belong. We are back with the dilemma. To free ourselves, we must look again at the relations between knowledge
and values, the criteria according to which we say something
is good, beautiful or true. The common statement that science
is value-free is wrong or, at least, so simplifies as to make very
little sense.

There is an alternative which, I think, is rightly called a
religious way out of the dilemma. This is to argue that we can
have knowledge of a purpose, plan, transcendent agency or *telos*
objectively existing in the world. Eighteenth- and nineteenth-
century philosophies of history took this view. There are
reasons or facts, it is said, for accepting the existence of a
purpose, even if they are not the reasons or facts of empirical
science. This, it scarcely needs saying, is an argument with a

long, inspired and richly diverse history from Plato to numerous modern scientists, to speak only of western thought. Other scientists and philosophers, however, deny such facts and reasons, and it is they who face the dilemma outlined. But the existence of the dilemma is, of course, no grounds at all for accepting a religious alternative; and I have not ventured into arguments for or against religious knowledge.

It is necessary to return to the anti-positivist reaction in the social sciences which, by the 1970s, had produced something like a new consensus: 'The new universe of discourse and sensibility . . . requires that we become increasingly aware that human beings are *self-interpreting* creatures, and that these interpretations are constitutive of what we are as human beings.'[1] This required new thought about objectivity and values in the human sciences, and the outcome is the belief that there are different forms of knowledge for different purposes. This conclusion sounds almost banal, because many people pay it lip-service; but, thought through, it has deep implications. One of them is that historical knowledge does indeed matter and should have authority in individual and collective self-knowledge. I support this conclusion through an analysis of the fact/value relation and discipline formation (since it is through the social institution of the discipline that values are embedded in modern scientific knowledge).

A categorical distinction between facts and values, understood as the distinction between 'is' and 'ought' statements, was a mainstay of the standard model in the social sciences. Time and again writers endorsed the judgement, dated back to Hume, that argument from 'is' to 'ought' involves an obvious fallacy. Twentieth-century positivists drew out the implications: whereas empirical statements may be true or false, 'value judgements are neither true nor false, may at most be regarded as expressions of certain psychological states, and may turn out to be meaningless exclamations'.[2] That is, according to this point of view, when we say that something is good or bad,

1 Bernstein, *Restructuring of Social and Political Theory*, p. 113. See Taylor, *Human Agency and Language*.
2 Kolakowski, *Positivist Philosophy*, p. 224.

beautiful or ugly, this is a verbal representation of a physiological, psychological or social state, itself describable in factual terms, or the expression is merely a habit which, though it may say something about ourselves or society, says nothing objective about the world. Many scientists found this philosophy congenial: it separated statements about facts and values, it refused to allow statements about the former to lead to statements about the latter, or vice versa, and it confined discussion about values to the social world outside the empirical sciences where it could be left to others. The rhetoric of the fact/value distinction was perhaps especially attractive to social scientists. It helped to demarcate new professional activity from the lay concern with social questions from which it had developed, and it drew a line between science and overt politics. For many natural and social scientists alike, it was precisely the ability of science to separate and keep apart facts and values, in contrast to other institutions which do not do so, which justified their claim to objectivity and cultural significance. In important respects, this remains the situation.

Yet, critics have for decades persuasively rejected the fact/value distinction. Here it is necessary to be clear. It is one thing to claim that scientists do not have the objectivity which they profess, that is, to claim that they are in some sense intentionally or unintentionally biased. It is another thing to claim that all work, scientific work included, is for a purpose or, as sociologists say, expresses an interest. This is not a claim about bias, though people often jump to the conclusion that it is; rather, it is part of a general theory of the conditions of all knowledge creation. As Weber stated: 'A chaos of "existential judgments" about countless individual events would be the only result of a serious attempt to analyze reality "without presuppositions" . . . Order is brought into this chaos only on the condition that in every case only a *part* of concrete reality is interesting and *significant* to us, because only it is related to the *cultural values* with which we approach reality.'[3] All statements of fact presuppose one kind of conceptual and theoretical scheme about the world rather than another, and the

3 Weber, '"Objectivity" in social science', p. 78.

facts can never fully determine which scheme to choose. This is a logical point not a point about bias. Its implication is that the use of one scheme rather than another depends on an evaluative judgement, an expression of 'interest', however buried in custom it may be.

There is a different kind of argument which states that values are themselves facts, that it is possible to give things like goodness objective, empirical description. Sherrington, who spent his career observing minute physiological facts, thought that the science of man, unlike the other sciences, must observe values. 'Nature becomes . . . a different object of contemplation for man if it faces him with a situation in which . . . Nature, in virtue of himself, has now entered on a stage when one at least of its growing points has started thinking in "values".'[4] This is very pertinent to the human sciences, since it is so common in liberal culture to characterise 'the human' in terms of inherent rights, freedom or dignity. There is a point of view in which these values are the *facts* on which human self-knowledge must build. I shall not, however, pursue this; it is enough to note that it makes the fact/value distinction highly questionable.

The most pertinent criticisms of the fact/value distinction can be made with the help of an example, one which was so influential that it added a new phrase to the English language. Dawkins's book on *The Selfish Gene* (1976) secured a large audience for new thinking about evolution by means of natural selection, and it pioneered a genre of highly successful popular science writing. There was some surprise when the book first appeared that it is indeed possible to write about technical matters in a way that both does justice to the complexities of science and grips the public imagination. The use of metaphor seems one clue to this achievement. As Vico observed, 'when people can form no idea of distant and unfamiliar things, they judge them by what is present and familiar'.[5] The metaphor of the book's title (about which, more in a minute) is obvious enough, but behind this lay a deeper

4 Sherrington, *Man on His Nature*, p. 5.
5 Vico, *New Science*, p. 76 (§ 122).

metaphor, a metaphor intrinsic to the intense appeal to Dawkins, and also it would seem to his many readers, of studying human evolution. 'We animals are the most complicated and perfectly-designed pieces of machinery in the known universe.'[6] People are people, who make machinery among other things; but here Dawkins stated that people are 'machinery' and, reinforcing the metaphor, stated that like machinery, they are 'designed'. As the finest machinery of all, people are the most fascinating thing to study. This is a clear statement of one 'conscious idea of man' (to repeat one of Berlin's phrases) underlying scientific knowledge.

Dawkins's use of metaphor was markedly ambiguous. On the one hand, he used metaphors as tropes, imaginative and literate ways to represent scientific claims in a form comprehensible to people with minimal technical background. This usage implied that it is in principle possible to discard metaphor and present each scientific statement non-metaphorically (e.g. in the mathematical notation of game theory, important to the modern account of natural selection). In such usage, the moral and emotional colour of words like 'selfish' and 'perfectly-designed . . . machinery' is adventitious, a way to arouse the reader's attention and not intrinsic to knowledge. In short, such usage hints that metaphor does not invalidate the fact/value distinction, even though the connotations of the metaphor may appear to do so. The facts, one might say, remain the facts, whatever the language used to describe them. On the other hand, Dawkins's use of metaphor was so omnipresent and so fertile that it is hard to imagine what would be left after an effort to strip it out. The metaphors appear to be the substance of what he had to say: humans really are 'perfectly-designed . . . machinery' and genes really are 'selfish'. Such usage clearly does not separate facts and values but rather identifies certain values, selfishness and perfectly designed machinery, as facts.

As empirical studies of scientific writing have shown and as post-positivist philosophers have argued, the distinction between literal and metaphorical description breaks down;

6 Dawkins, *The Selfish Gene*, new preface, p. vi.

metaphor is intrinsic to all thought about the world. 'Every word or phrase has its contexts or language games in which usage is recognized as normal and in which a standard meaning can be invoked, and others in which the use is unfamiliar or novel. There is no rigid distinction between the literal and the metaphorical.'[7] The distinction boils down to a distinction between familiarity and unfamiliarity. The language of professional science is indeed full of statements which began life as clear metaphors.[8] And when colourful metaphors appear in writing for a general audience, the connotations that language has, and which are often hidden (but not lost) in the dry, conventional format of science writing for specialists, reappear. 'As soon as scientific writers move across levels of language and reference, the apparent autonomy and neutrality of description are shaken. One might add that the shared assumptions of the group begin to be visible.'[9]

Dawkins could not have presented his account of the genes as a list of non-metaphorical and empirical claims assembled without a plot. There is, necessarily, more to exposition than this. One suspects, however, that he tried 'to have his cake and eat it': he leant on the empirical authority of science, based on its reputation that it separates values from facts and thus achieves objectivity, and at the same time he used language which encouraged readers to draw evaluative conclusions from facts. In this regard, the phrase 'the selfish gene' was a masterstroke. Given common language use, Dawkins's conjunction of these two words richly conveyed belief, here presented as scientific knowledge, that the most elementary stuff of life determines that self-interested actions, originating in individuals, shape the human sphere. As a matter of explicit statement, Dawkins neither owned nor disowned such a view. All the same, the text that readers read rendered the values of competitive individualism as the expression of nature. The metaphor carried the content of the message. As Gadamer wrote, 'to regard the metaphorical use of a word as not its

7 Arbib and Hesse, *Construction of Reality*, p. 145.
8 For a case study, see Smith, *Inhibition*.
9 Beer, 'Problems of description', p. 45.

real sense is the prejudice of a theory of logic that is alien to language'.[10]

The point is not that Dawkins's book was wrong to make assumptions about human nature. I have repeatedly stressed that any science of human nature makes assumptions about what a human being is. Two other things are wrong with Dawkins's argument. Firstly, he used the rhetoric of empirical science to imply that he made no such assumptions, but built only on evidence. Secondly, the theory of human nature which he did choose, competitive individualism, has many moral and political flaws. By saying this, naturally, I take a political stance; but so did Dawkins, whether that was what he intended to do or not. Prior commitments were – as they must be for all writers – built into the language. Glib remarks about value-free science take no account of this.

There is still another philosophical point to make. Words, including 'gene' as well as 'selfish', have meaning not in themselves but as part of a narrative about the world (or even as part of a narrative about a narrative about the world, and so on). These narratives are historically and socially created. 'Gene' denotes a material pattern of chemicals with a place in organising what an individual living organism becomes; and 'selfish' refers to the actions of a social person judged by a moral standard. 'Gene' normally belongs in narratives which scientists write about material inheritance and the relation of that inheritance to an organism's development. 'Selfish' normally belongs in narratives about morality, justice and the sheer business of people getting on together. Take away the communities which, over historical time, have built up the narratives of which the words are part, and *neither* word can be said to mean anything. This is easier to see in the case of 'selfish'; after all, we constantly argue and negotiate about descriptions of selfishness. But the principle is just the same in the case of 'gene'. It is the historical achievement of communities of scientists to have picked out particular parts of nature, to have called them 'genes' and to have made that description stick. But no gene is or could be an isolated unit,

10 Gadamer, *Truth and Method*, p. 429.

an independent entity with a causal power of its own. The word 'gene' helps represent a fabulously complex process in terms of comprehensible and manageable parts.

Meaning belongs to language-using communities not to nature. It is this, at base, which puts paid to attempts to find in nature criteria with which to judge what is good or bad in human life. To make this clear, one can recall what the physical chemist, economist and philosopher Michael Polanyi called the Laplacean fallacy. The French mathematician Laplace, who in important respects filled out Newton's mechanistic view of the universe, imagined a mind with complete knowledge of the position and motion of all particles of matter at any one moment. This super-mind, Laplace said, would be able to know everything that has happened in the past and will happen in the future. It was a marvellous image of the power and scope of mathematical mechanics. The trouble is, as Polanyi observed, that 'the Laplacean mind understands precisely nothing and that whatever it knows means precisely nothing'.[11] The super-mind has a list of mathematical points and equations and rules for calculations; the human mind, however, embeds all this in a meaningful set of claims about the world – it has *knowledge*. The Laplacean mind is not, after all, a *mind*.

It is worth pausing with the fantasy of a non-metaphorical and purely factual description of the world. What could this be like? Description immediately grinds to a halt. 'Out there' (but this is a metaphor) there is stuff (another metaphor), or processes (a juridical metaphor), or whatever, carrying on. There is no meaning, pattern or purpose; there is no 'earth' and no 'life', and there are no 'genes' and no 'people'. Wittgenstein's statement, made in his early positivist period, retains its force in relation to any claim to describe the world independently of human values: 'The sense of the world must lie outside the world. In the world everything is as it is, and everything happens as it does happen: *in* it no value exists – and if it did exist, it would have no value.'[12]

---

11 Polanyi, *Personal Knowledge*, p. 141.
12 Wittgenstein, *Tractatus Logico-Philosophicus*, § 6.41.

These arguments lead, I think, to the conclusion that all language is metaphorical. Nietzsche certainly thought this. But it is an argument into which it is not necessary to go further.

Dawkins took two words, 'gene' and 'selfish', treated them as if they were descriptions of things or states existing independently of how people use words, and then combined the words which had previously belonged to two unrelated narratives. He thereby set about bringing into being a new kind of world, one in which it is indeed possible, and meaningful, to say that genes are selfish. He seems to have had some success. Whether this is a world in which we all want to live is another matter. Genes are elements of complex processes, picked out as units for purposes of clarity in ordering research, in stating knowledge and in the hope of relating the observable characteristics of different individuals and groups to underlying causes, thereby opening up possibilities for manipulation. Genes, like the causes attributed in a criminal trial, are part of a story told for a purpose. To pick out the gene as the star of the story and then call the gene 'selfish' tells one story rather than another. Whatever the scientific authority of such stories, they incorporate judgements about what is good or bad in their conceptual language and narrative form. Marx made just this point when he criticised Feuerbach for taking the material world as something given, not made: 'the sensuous world around him is, not a thing given direct from all eternity, ever the same, but the product of industry and of the state of society'.[13] A scientist like Dawkins sought entities, such as genes, which could serve as the objective foundation for an understanding of what is human, but what he wrote about these entities was at least in part 'the product of industry and of the state of society'.

The discussion of narrative reached similar conclusions: no narrative is neutral, and many narratives are possible about any event. It cannot be said that objects in the world fully determine what we say about them. Geertz made the point with special clarity in relation to the writing of ethnologists:

13 Marx and Engels, *The German Ideology*, p. 35.

'The burden of authorship cannot be evaded ... [There is a] strange idea that reality has an idiom in which it prefers to be described, that its very nature demands we talk about it without fuss – a spade is a spade, a rose is a rose – on pain of illusion, trumpery, and self-bewitchment.' – '[But] descriptions are homemade ... they are the describer's descriptions, not those of the described.'[14] As he also noted, however, this certainly does not make our belief that we can describe facts, and judge whether a description is a good one, into a delusion. Scientists have elaborate ways of doing just these things.

## Truth as a value

It is an impressive achievement to have established belief that only one kind of community, a scientific community, has the means and the authority to deliver true descriptions. There are significant settings where this authority does not hold – in legal decision-making, in religious societies and in politics generally – but it is sufficiently strong to have created, in the West, large areas of life in which other forms of knowledge have to measure themselves against scientific knowledge, while scientific knowledge itself is thought to be self-confirming. Explaining this has preoccupied historians of science. Yet, I am now envisaging the existence of different forms of knowledge for different purposes.

This step comes right up against the straightforward commitment scientists and ordinary people alike have to the truth. Science is significant precisely for this commitment, in contrast with other areas of life, like politics, where truth may not be a consideration. To be a scientist is *to profess a way of life* which expresses the value of truth, and whenever a scientist proffers a claim held to be true, he or she voices this value. Any particular claim may be false, or at least uncertain, but the act of making the claim still upholds the principle. Though truth is a self-evident good for scientists living in a scientific culture, 'the very definition of "true" knowledge rests in the final analysis upon an ethical postulate', as the molecular

14 Geertz, *Works and Lives*, pp. 140, 145.

biologist Jacques Monod observed.[15] This 'ethical postulate' is not something which scientific knowledge itself gives; it comes from the development of a culture, and it could indeed disappear if that culture were to disappear. (The fear that this might happen informed Polanyi's work on the philosophy of science.)

Nietzsche's reputation rests in some large part on his having perceived that truth as a goal is a cultural value which, like other values, is open to critique. 'The will to truth requires a critique – let us thus define our own task – the value of truth must for once be experimentally *called into question*.'[16] Many of Nietzsche's academic contemporaries feared that an all too real, rather than 'experimental', questioning of the value of truth was under way, and they blamed both the positivist philosophy which they attributed to natural scientists and some historians, as well as Nietzsche himself. Some modern scientists perhaps feel the same about sociologists of scientific knowledge. In the midst of the earlier 'crisis', however, Weber made lastingly clear that science begs the question of its own fundamental value. His statement about the unsecured purpose of science as the pursuit of truth has acquired canonical status: 'Tolstoy gave the simplest answer to the only important question: "What should we do? How should we live?" The fact that science does not give us this answer is completely undeniable. The only question is in what sense does science give us "no" answer and whether or not it could perhaps be of use to somebody who poses the question properly.'[17] Nietzsche's statement was as emphatic: 'Science is not nearly self-reliant to be that [the alternative to faith in the ascetic, in God]; it first requires in every respect an ideal of value, a value-creating power, in the *service* of which it could *believe* in itself – it never creates values.'[18] If we suppose that it makes sense to seek the truth, we cannot evade Heidegger's questions: '*Why must we presuppose that there is truth*? What

15 Monod, *Chance and Necessity*, p. 161 (italics removed).
16 Nietzsche, *Genealogy of Morals*, p. 153.
17 Weber, 'Science as a vocation', p. 18.
18 Nietzsche, *Genealogy of Morals*, p. 153.

is "presupposing"? What do we have in mind with the "must" and the "we"?' His response was that 'it is not we who presuppose "truth"; but it is *truth* that makes it at all possible onto-logically for us to be able to *be* such that we "presuppose" anything at all'.[19] The alternative to this turning inwards to ontology, an alternative now widely canvassed outside the natural sciences and religious societies, and associated espe-cially with the work of Rorty, is to discard all this talk about truth.[20]

Scientists work in specialist communities of like-minded people who together defend the value of truth-telling among themselves and do not usually go into questions of ontology. Scientists, especially social scientists, however, do not limit the value they attach to truth-telling to their own commun-ities but frequently extend it as an ideal to humankind as a whole. This is the Enlightenment ideal. Most social scientists 'have assumed that "enlightened" men share the same basic values, and that the important task is to gain a fuller empir-ical understanding of the consequences of possible sources of action, and the empirical means to foster those values which enlightened men endorse'. But, as Richard J. Bernstein, who made this comment, went on to say: 'They blur and gloss over the abyss that Nietzsche had uncovered and Weber looked into – that there can be no ultimate rational foundations for our basic values.'[21] Scientific truth-telling has not provided rational grounds for itself, let alone for extending that prac-tice to human affairs. Bernstein wrote in the 1970s, at the end of several decades of rapid expansion of the psychological and social sciences, funded with the expectation that true know-ledge of human beings would help usher in a rational world. Scientists then felt little need to question basic enlightened values. But this has changed as the world around has changed. The search for truth, as the discussion of reflexivity suggested, has led to some searching questions about the philosophical as well as practical limits to reason.

19 Heidegger, *Being and Time*, p. 270.
20 See Rorty, *Contingency, Irony, and Solidarity*.
21 Bernstein, *Restructuring of Social and Political Theory*, p. 49.

Weber himself discussed the problem of 'ultimate rational foundations' in the tragic mode appropriate to a problem which we must live with rather than solve. A famous passage is one of the defining descriptions of modernity:

> Only positive religions – or more precisely expressed: dogmatically bound *sects* – are able to confer on the content of *cultural values* the status of unconditionally valid *ethical* imperatives . . . The fate of an epoch which has eaten of the tree of knowledge is that it must know that we cannot learn the *meaning* of the world from the results of its analysis, be it ever so perfect; it must rather be in a position to create this meaning itself. It must recognize that general views of life and the universe can never be the products of increasing empirical knowledge, and that the highest ideals, which move us most forcefully, are always formed only in the struggle with other ideals which are just as sacred to others as ours are to us.[22]

This leaves to social scientists relatively modest tasks, such as classifying different kinds of social action; they cannot formulate or decide between 'ideals' in the way someone like Rickert thought the business of the true scholar. In Weber's analysis of social action, action is of different kinds because there are different values and purposes; different social phenomena have different 'value-relevance'.[23] His own research was on what people take 'value-relevance' to be, not on justifying any particular value. Scientists, for example, judge the 'value-relevance' of a topic from its perceived possibilities in adding to truth in ways which the scientific community (in fact, a very small specialised part of it) will recognise. He himself distinguished four types of action in which people have an interest: instrumental rational, value-rational, affectual and traditional. The first involves rationally calculated ends (e.g. in engineering), the second pursues an ideal (e.g. a scientist pursuing truth), the third follows feelings and bodily states, and the last are simply habit and custom.[24]

22 Weber, '"Objectivity" in social science', p. 57.
23 Weber, 'Meaning of "ethical neutrality"', p. 21.
24 Weber, *Economy and Society*, vol. 1, pp. 24–6.

Responses to the 'abyss' into which Weber looked have diverged. The consensus among natural and social scientists alike for many decades was not to look over the edge but to separate facts and values. This, they assumed, left them free to research the facts while handing values over either to professional philosophers in ethics, political theory or aesthetics, or, when decisions were needed, to the reigning powers. This – for some – comfortable arrangement has now broken down, faced, for example, by genetic engineering of new life-forms, though its rhetoric persists. It also broke down because it was self-contradictory. As the discussion of 'the selfish gene' suggested, scientists actually wanted their bread buttered on both sides. They asserted the objectivity of a value-free account of nature (as in the detail of Dawkins's book) and asserted values (such as the primacy of scientific reason in human affairs, or, as in Dawkins's title, a much more direct view of what nature has to teach). Writing about the social sciences, Bernstein observed: 'despite all the talk of objectivity and value-neutrality, social science literature and so-called empirical theory are shot through with explicit and implicit value judgments, and controversial normative and ideological claims'.[25] Despite the rhetoric of the fact/value distinction, the literature on sociobiology, neuroscience and evolutionary psychology, like the literature on 'the selfish gene', is also 'shot through' with 'value judgments, and controversial normative and ideological claims'.

A number of highly successful science writers try to have it both ways: to claim that evolutionary or brain science is exemplary in its objectivity, and to claim that science reveals human nature, how people necessarily – at base – feel and act, and hence that nature is normative for human nature.[26] The objection is not in itself that such writers presuppose norms, but that they do not allow, or they at least implicitly denigrate, other forms of knowledge about being human which might support other norms. In particular, they do not allow knowledge about their own knowledge as historically created and historically situated activity. Where writers take nature as

25 Bernstein, *Restructuring of Social and Political Theory*, pp. 52–3.
26 See Farber, *Temptations of Evolutionary Ethics*.

guide, as in the very notion of 'the selfish gene', they dismiss the knowledge that we have that norms are in their nature social. In fact, '"oughts" exist only for members of groups' and '"obligation" and "ought" presuppose the existence of social institutions, and it is impossible to reduce their meaning to non-social, say biological or psychological, categories'.[27] Belief that it might be possible to make a statement about what it is right to do independently of the social world in which the notion of something 'being right' has meaning is itself a judgement in need of historical explanation.

In any case, the truth, utility or profit that scientists pursue is not something that exists in, or something that can be extracted from, a world 'out there' independently of how scientists, in historical fact, shape their knowledge of that world. Scientists who find such values in nature implicitly make assumptions with a religious or metaphysical character. These assumptions are of the kind Whitehead had in mind when he wrote: 'At the base of our existence is the sense of "worth" . . . It is the sense of existence for its own sake, of existence which is its own justification, of existence with its own character.'[28] Certain values, which Whitehead here called 'worth', appear to be given with the fact of existence, and these values scientists often enough project into nature, as if they were 'in nature', as scientists themselves represent nature, as opposed to 'in existence'. Whatever we feel and say about such 'worth' (this is a metaphysical or religious matter), however, it can never lose its historically embedded character. A number of modern philosophers indeed, like MacIntyre, have argued that such a sense of 'worth' can make sense only in the light of a tradition.

The rational standards which, scientists believe, give, or should give, scientific knowledge incomparable authority require values, convictions about what constitutes human flourishing or 'worth'. 'Any choice of a conceptual scheme presupposes values . . . One cannot choose a scheme which simply "copies" the facts, because *no* conceptual scheme is a

27 Kusch, *Psychological Knowledge*, pp. 255, 325.
28 Whitehead, *Modes of Thought*, p. 149.

mere "copy" of the world. The notion of truth itself depends for its content on our standards of rational acceptability, and these in turn rest on and presuppose our values.'[29] There is, therefore, need for historical knowledge of where our values come from, and this holds for scientific as well as ethical values. 'Our views of the nature of coherence and simplicity are historically conditioned, just as our views on the nature of justice or goodness are.'[30] In modern science, the most important site of 'historical conditioning' is the discipline or sub-discipline in which a scientist trains and works.

### Disciplinary knowledge

Modern disciplines carve up intellectual life: 'disciplines are . . . the primary unit of internal differentiation of the modern system of science'.[31] Within disciplines, or sub-disciplines, something like a one-dimensional view of the world can and does reign. Scientists and scholars perceive and judge the world, at least when they are being 'professional', in terms of the form of knowledge appropriate for the purposes of the academic or technical occupational group to which they belong.

Precisely because of this, discipline hybridisation (for example, the movement of physicists into molecular biology after 1945) and 'boundary work' policing or subverting institutionalised practices is important to innovation. Funding agencies, following a business model, have made considerable efforts to promote project-oriented, rather than discipline-oriented, research. There are important precedents in the human sciences: in the 1930s, the Rockefeller Foundation hoped to synthesise the social and psychological sciences at the Yale University Institute of Human Relations, and in the 1950s the Ford Foundation supported similar goals under the heading of the behavioural sciences. More recently, researchers have turned

---

29 Putnam, *Reason, Truth and History*, p. 215. Also Putnam, *Collapse of the Fact/Value Dichotomy*.
30 Putnam, 'Beyond the fact/value dichotomy', p. 138.
31 Stichweh, 'Sociology of scientific disciplines', p. 4 (italics removed). See Ross, 'Changing contours of the social science disciplines'.

to inter-disciplinary, cross-disciplinary or trans-disciplinary perspectives in response to what they regard as the constraints imposed by existing institutions. Social psychology, sand-wiched between the psychological and social sciences, is in a particularly sensitive position.[32] Yet, in spite of all these experiences and initiatives, power remains with the socially embedded disciplines, as the confusing multiplicity of words used in the attempt to lever open new institutional possibilities perhaps indicates.

It requires an act of imagination to shape research on other than disciplinary lines. This is one reason why the term 'the human sciences' is useful. It does not describe a discipline but creates social space where disciplines seek to co-operate. It allows for interaction among the different forms of know-ledge of being human across the full range of the sciences (in the continental European sense).

The institutional structures of nearly all disciplines are rela-tively modern. In the case of disciplines in the human sciences – like political science, economics, social anthropology, lin-guistics, geography and sociology – they date from the late nineteenth century. Substantial expansion took place mainly in the mid-twentieth century. The disciplines in the natural sciences developed somewhat earlier, and physics, chemistry, physiology, geology, zoology and botany all date from the first half of the nineteenth century. The disciplines of the Anglo-American humanities faculty – philosophy, history, Classics, literature, languages and the arts – developed in complex ways over a substantial period of time. But the recognisably modern profession of history, for example, also dates from the early nineteenth century, when German scholars built up the prac-tice of training with primary source documents in the research seminar. The academic disciplines related to the three medi-eval professions – law, medicine and theology – have more obvious continuity with earlier centuries.

A number of people have suggested that something very important happened to knowledge with the development of

---

32 Good, 'Quest for interdisciplinarity'; Good, 'Disciplining social psychology'.

modern disciplines. Thomas Kuhn, for example, referred to 'a second scientific revolution' in the first half of the nineteenth century.[33] Knowledge acquired a self-perpetuating disciplinary character: 'scientific disciplines now represented real systems organizing themselves'.[34] One consequence was that knowledge not constituted in an institutionalised discipline lost status as knowledge, and terms like 'folk knowledge', 'eclectic' and 'amateur' acquired their present negative connotations. The introduction of a definite pattern of training in research institutions transformed knowledge. 'Thus the transition took place from relatively flexible intellectual genres to far more strictly organized university disciplines. This transition meant not only more specialization, but also a more constraining type of organization.'[35] The sheer quantity of research and knowledge grew enormously, and with it specialisation; all the same, specialisation was a social process and not, in a literal sense, a natural outcome of the growth of knowledge. Specialisation built disciplinary authority into institutions, into the knowledge and expertise of well-defined groups of people and social structures. In particular, natural scientists began to write exclusively for an audience sharing special training and practice. This established well-defined scientific communities with certain standards of (in Putnam's words) 'rational acceptability'.

The change from seventeenth- and eighteenth-century natural philosophy to nineteenth-century natural science shaped a number of specialised preserves out of general views of nature. There was a parallel change in moral philosophy when scholars shaped the specialist disciplines of philosophy, political economy, history and philology. These steps altered possibilities for creating meaning and significance in human life and were not just adjustments in the manner of organising knowledge. The Hungarian social philosopher Gyorgy Markus pointed out, for example, how the professional commitment to the depersonalised and specialised scientific paper narrowed

33 Kuhn, 'Function of measurement', p. 147; Kuhn, 'Relations between history and history of science', p. 220.
34 Stichweh, 'Sociology of scientific disciplines', p. 7.
35 Heilbron, *Rise of Social Theory*, p. 3.

the cultural significance of the natural sciences at the same time as it raised their technical effectiveness. Papers became not just difficult to read but incomprehensible without training in, and experience of, the practice to which they were a contribution. The disciplining of scientists ensured that 'their discourse became self-enclosed (i.e. one among experts alone) in the very same process in which they *acquired* those epistemic and social characteristics which made them able to fulfill a direct function in technical development'.[36] The sciences formed into specialist areas and thereby secured the means to achieve progress as it was understood within those areas, but they thereby lost the claim once made for knowledge, that it could secure a meaning for life at large. This was equally the case in the social sciences and in the natural sciences. Social scientists have even described a recurring inverse correlation between technical precision in a domain and the domain's public meaning or significance.[37]

Specialisation also involved a process of deracination. The training of specialists introduces them to one discipline's notions and methods of understanding. Appreciation of the historical roots of specialist activity at a time when there was no such specialisation, or of the wider contemporary social meaning of specialist knowledge, requires different skills and knowledge, which teachers rarely attempt to provide. Indeed, scientific specialists systematically, implicitly if not explicitly, teach their students, especially through rigorous courses on experimental technique or methodology, that specialist knowledge is independent of social context, past or present. The emphasis on techniques and methods has the effect of making objectivity appear something which individuals may acquire, like a skill. All the same, objectivity is a set of social rules, or standards, applied in a particular way, not the accomplishment of individuals, however gifted, establishing an unmediated relationship with what they think is real.

---

36 Markus, 'Why is there no hermeneutics of natural sciences?', p. 45.
37 For example, the Dutch psychologist Johan T. Barendregt's 'neurotic paradox'; see Dehue, *Changing the Rules*, pp. 93–4, 112–19.

The formation of large and, at times, extremely well-funded occupational groups of scientists has institutionalised particular ways of life with particular purposes. These ways of life undoubtedly make rigour and agreement in stating knowledge possible. Within scientific groups, there are times when something like an ideal form of community, with shared standards, practices, knowledge and goals, prevails. When this occurs, those who experience it may feel that it flows from the group's ability to mirror nature, and other people may be sufficiently impressed to agree. But agreement is a social achievement. This can be seen clearly when knowledge in the domain changes or when the group deploys its knowledge for purposes which the group itself does not define. *Outside* the group different purposes and standards often exist, and where that is the case, a larger consensus is much more difficult to achieve. This is visible, for example, in legal contexts. Parties to court cases may call on scientific experts, and it is notorious in Anglo-American courts that both prosecution, or litigant, and defendant, or respondent, will find experts who disagree with each other. Experts have been heard to complain that courts are not concerned with the truth; and indeed they may not be concerned with the truth as a scientist's own occupational group understands it. The courts, however, have other purposes and the power to impose other purposes on scientists who appear before them. Which purpose? Whose truth?

We might compare discipline formation with the creation of the nation state: both historical processes have given rise to basic structures in terms of which we describe social reality. Both disciplines and nations give an apparently natural and unquestionable identity to their members. Nonetheless, disciplines and nation states are modern, historically contingent ways of shaping what people hold to be right and true.[38]

## Different forms of knowledge

Modern claims to the exclusive objectivity and truth of natural science knowledge of human nature, as well as dogmatic

38 Blondiaux and Richard, 'A quoi sert l'histoire des sciences de l'homme?', p. 117.

religious, ethnic or other kinds of claims, though they contradict each other, are 'dead set' against anything hinting at relativism. As a philosophical problem the rationality/relativism issue has its place, and a lot of ink has been spilt in connection with it. But no arguments made here make a philosophical case for some kind of absolute rationality or relativism (whatever 'absolute' might mean) in knowledge or values. There is no need. There is more than enough to do to create persuasive arguments, arguments worthy of consideration *in the context* of what now has significance. There is much more to be gained, for most purposes, from getting on with these arguments than prejudging whether they lead to rationality or relativism, which, after all, may be a false antithesis. The arguments made about the reflexive nature of knowledge support this: we cannot state a position without some presuppositions, and we must, as a consequence, get on with the argument in hand. Standards of objective argument and persuasiveness operate in all the sciences; there is certainly enough knowledge of what is objective and what not for present purposes.[39]

The point is muddied because of the assumption that 'the truth' is unified, of a piece. This itself is a value judgement: there may be different kinds of truths. 'The privileged status of scientific rationality . . . raises the problem of whether there is a plurality of irreducibly different but equally legitimate criteria for rationality, each appropriate to its own sphere.' I am now arguing that there are. The reason is, to continue in Oakes's words, that the 'differences between the sciences are a result of differences in their distinctive theoretical interests or purposes'.[40] As I stressed earlier, it is a misunderstanding to think that it is subject matter which determines the classification of knowledge. What justifies distinctions between the natural sciences and the human sciences (here including both social sciences and humanities) is the concentration, though

---

39  See Megill (ed.), *Rethinking Objectivity*; Nelson, Megill and McCloskey (eds), *Rhetoric of the Human Sciences*; Barbara Herrnstein Smith, *Belief and Resistance*.
40  Oakes, *Weber and Rickert*, pp. 2, 22.

definitely not exclusive presence, in the different sciences of different forms of knowledge for different purposes.

Let me take as an illustration the way a number of popular science writers, like Pinker, have described the triumphant advance of evolutionary science and neuroscience to encompass human nature.[41] They portray natural science taking over territory previously closed to it because people, out of ignorance or prejudice, believed that part of existence, the soul, was not accessible to science. Hence the colossal emphasis on Darwin, whose evolutionary theory demonstrated both the empirical evidence for 'man's place in nature' and explained how it came about. Darwin, certainly, did have opponents, and even some colleagues, who argued that there are parts of the world to which the natural sciences do not apply. Asa Gray, for example, the Harvard botanist who was a loyal friend to Darwin, thought that there is a guiding power at work in causing intra-specific variation. Such argument offered a hostage to fortune since, in principle, in as far as anything is part of the world it is a possible object of scientific knowledge; and later scientists indeed explained variation in natural terms. The position that Gray took, however, is not the profound point at issue in resistance to the picture of natural science as all-conquering. The deep point at issue is whether there are other purposes than those which natural scientists pursue and, if so, whether these other purposes make other forms of *scientific* knowledge not only possible but necessary.

Many people would, I suppose, think that the natural sciences make progress and advance knowledge in a way that the social sciences and humanities do not. This was indeed a claim in the influential founding statement of evolutionary psychology, 'The psychological foundations of culture' (1992), by Tooby and Cosmides. The judgement that the natural sciences make progress and the social sciences and humanities do not implies that any activity claiming to be a human science should become a natural science in order also to achieve the same self-evidently desirable goal. It is a forceful point for politicians and the public, who fund scholarship, and for

41 Pinker, *How the Mind Works*.

administrators and scientists themselves, trying to distinguish good and less good (or just plain bad) research. The criterion of progress – Is research likely to increase knowledge? – seems the obvious one to take.

The goal of evolutionary psychology is to explain universal and durable human psychological characteristics as the result of selection pressures operating during the early period of human evolution. My concern, however, is with the proponents' indictment of the social sciences for lack of progress and their claim that only when these sciences integrate with biology will they achieve it. 'We suggest that this lack of progress . . . has been caused by the failure of the social sciences to explore or accept their logical connections to the rest of the body of science – that is, to causally locate their objects of study inside the larger network of scientific knowledge.' The principal symptom of this 'failure', in the authors' view, is the abandonment of causal explanation for the study of texts, freeing 'scholars from all of the arduous tasks inherent in the attempt to produce scientifically valid knowledge'.[42] The judgements about the virtues of being 'arduous' and that one kind of scholarly work but not another is indeed 'arduous' are worth noting as examples of the evaluative discourse that does indeed appear in scientific papers. But it is the point about progress which matters. While Tooby and Cosmides stated that 'by calling for conceptual integration in the behavioral and social sciences we are neither calling for reductionism nor for the conquest and assimilation of one field by another', they nevertheless argued that only the kind of evolutionary psychology which they proposed makes it possible to formulate testable hypotheses, and hence make progress, in the social sciences.[43] There is evidence in their favour. As Markus, no sympathiser with such opinions, observed: 'Whatever one's view of the idea of a unilinear scientific progress, it is the modern natural sciences which indubitably provide at least

---

42 Tooby and Cosmides, 'Psychological foundations of culture', pp. 22, 23.
43 Cosmides, Tooby and Barkow, 'Introduction: evolutionary psychology and conceptual integration', p. 12.

the best approximation to what should be understood by the notion of an "accumulative historical growth" – the process of continuous *tradition-transmission* and simultaneously creative and accretive transformation of this tradition.'[44] This is widely agreed, so much so that many people, if asked to say what progress means, might point to the 'accretive' tradition of natural science.

There are a number of possible responses to Tooby and Cosmides. The first is to suggest that the differences between the natural and social sciences may not, after all, be so great. Progress is not a feature of science in general, something that has no concrete existence, but of particular areas of research, each of which has somewhat different criteria for assessing what counts as progress. Each discipline or sub-discipline, including history and the history of science, has standards for judging whether one piece of knowledge marks an advance over another. These standards may, at times, be much less clear cut in the human sciences than in the natural sciences, but the fact that one interpretation rather than another gains ascendancy in a field shows a kind of progress. For the few specialists in the history of psychology, for example, there have been considerable advances over earlier statements that 1879 is the founding date of experimental psychology. Conversely, there are areas of the natural sciences where progress is debatable, very slow or even non-existent; perhaps it is not overly tendentious to cite cancer research as a case in point (but even here the field is too large sensibly to talk about progress, or lack of it, in general).

It is also important to note the social fact that progress, whatever else it may be, is measured in terms of funding allocations, citation counting and career advancement. There is no question that the bureaucratic mind has found it much easier to put these measures into effect in the area of the natural sciences and empirical social sciences than in the area of the humanities. Thus, were it to be the case that one research programme along the lines proposed for evolutionary psychology were to achieve dominance in the social and

44  Markus, 'Why is there no hermeneutics of natural sciences?' p. 9.

cultural sciences, it would set out a bureaucratically attract-ive, level playing field on which researchers might compete for funds. Viewed in this context, advocacy of one kind of progress amounts to an attempt to consolidate one form of rationality, a politically and bureaucratically instrumental one, as rationality in general. There are strong rational and polit-ical grounds for objecting to this. 'Given the socio-cultural preconditions of modernity, natural science is an intellectual enterprise with the inherent ability to "progress," but any attempt at the definition of the criteria of this progress with-in some framework independent from transient historical-cultural variables seems . . . doomed to failure and leading only to the hypostasis of some particular cultural characteristics as universal constituents of human rationality.'[45] What Tooby and Cosmides denoted by 'progress' is not what others mean by the word. They used a word taken from the wider culture without the tools for an effective critique of what they were doing. Progress in critique of words like 'progress' is precisely what a different view of the social sciences might have pro-vided them with. We can understand what progress means within a particular research programme clearly enough, but whether that is the same as progress in a larger sense is another matter.

It was something of a cliché of twentieth-century social commentary to point to a gap between progress in the physical sciences and technology and progress in human affairs. Dewey, for example, referred to the fact that 'the need of . . . control at the present time is tremendously accentuated by the enorm-ous lack of balance between existing methods of physical and social direction'.[46] Raymond B. Fosdick, President of the Rockefeller Foundation, reorganised in 1928, used the notion of a 'lag' to support a systematic effort to create a science of man.[47] He believed that with appropriate political and finan-cial commitment the social sciences would catch up with the natural sciences, and control over human affairs would catch

45 *Ibid.*, p. 45.
46 Dewey, 'Need for social psychology', p. 275.
47 Kay, 'Rethinking institutions'.

up with control over nature. This did not happen, at least not on the large scale and in the way that promoters of social science like the Rockefeller and liberal-democratic governments envisaged. The social sciences did not live up to the hopes once invested in them, and at the end of the twentieth century it was a politically divisive matter to say why. One group of people took the view that the social sciences had failed to make progress because they had not become objective and scientific. Tooby and Cosmides suggested a corrective. Another group of people, however, thought that unquestioning commitment to one kind of science was itself the problem. In the judgement of the historian of science, Steven Shapin, for example, writing about the long-term implications of the seventeenth-century scientific revolution, 'good order and certainty in [natural] science have been produced at the price of disorder and uncertainty elsewhere in our culture'.[48] From this perspective, no amount of-natural science, however authoritative, can address social uncertainties, and it makes good sense to promote the social or human sciences as disciplines suited to examine the nature of progress. This point of view is needed to understand the oft-expressed, though perhaps naively formulated, judgement that scientific advance does not deliver 'real' progress, that is, progress in the moral or spiritual quality of life.

The supposed lag between the natural sciences and human affairs may be no lag at all but a gap between different purposes and forms of life. There is no catching up to be done. In so far as the natural sciences, the social sciences and humanities have different purposes, there can be no common criterion of progress, and any concerted effort to impose one should be seen for what it is – a political move to expand one form of life at the expense of another.

What, then, are the purposes of the kind of human sciences, including history, which I have been arguing for? To start, we must return to an argument such as Gadamer's: the humanities (Gadamer's translator used 'human sciences' for '*Geisteswissenschaften*'), and the social sciences to the extent to which they belong with the humanities, are not characterised by

48 Shapin, *The Scientific Revolution*, p. 164.

progress, measured in terms of knowing and manipulating an 'object in itself' but express 'life itself'. Gadamer defended the human sciences (or humanities) since they create 'knowledge of another kind and order' to the physical (or social) sciences.[49] Discussing historical knowledge, he wrote: 'The theme and object of research are actually constituted by the motivation of the inquiry. Hence historical research is carried along by the historical movement of life itself and cannot be understood teleologically in terms of the object into which it is inquiring. Such an "object in itself" clearly does not exist at all.'[50] If the object of knowledge is not, as we might say, 'out there' but *is* reflexive interpretation, then there is not and cannot be progress, in this realm, of the kind found in the natural sciences or technology. Nevertheless, there assuredly have been and are ways of evaluating good and bad arguments in the disciplines of reflexive knowledge. In addition, one of the activities of reflexive knowledge is to seek to determine what is 'left out' by understanding the world in terms of one language, including the language of rational science, and one way of life rather than another.

A thing, person or event has value or significance by virtue of its place in a narrative re-creation of the world. 'Progress' means something to evolutionary biologists because they tell a story about how knowledge in their field has expanded its scope and detail, how knowledge brings intellectual satisfaction and material rewards and how this field fits into a larger picture of western intellectual history. Other people, however, tell stories about how this knowledge has not brought happiness, or justice, or well-being, or a sense of a secure future, or spiritual or moral enrichment. There is no putting forward a judgement about what progress or any other value is independently of the narrative that makes the judgement a meaningful one. Narratives are in their nature historical, and hence historical knowledge is implied, even if it is not made explicit, in any judgement, in science as elsewhere. 'An understanding of the concept of the superiority of one physical [or other

49 Gadamer, 'Truth in the human sciences', p. 26.
50 Gadamer, *Truth and Method*, pp. 284–5.

scientific] theory to another requires a prior understanding of the concept of the superiority of one historical narrative to another. The theory of scientific rationality has to be embedded in a philosophy of history.'[51] The historical narrative which makes sense of a judgement about progress in the physical sciences is different from the narrative which makes sense of a judgement about progress in the human sciences. Not the least of the reasons is the fact that the human sciences include assessment of just the kind of argument I am now making, that is, their subject matter is reflexive.

We have reached a decisive rational argument for historical knowledge: it is a condition of knowing what we mean by statements. Normally, historical knowledge is simply taken for granted; all disciplines equip their students with a wealth of tacit historical knowledge. The business of the discipline of history is to make the knowledge explicit, and this business is intrinsic to science not something apart from it. 'Any attempt to show the rationality of science, once and for all, by providing a rationally justifiable set of rules for linking observations and generalizations break down ... It is only when theories are located in history, when we view the demands for justification in highly particular contexts of a historical kind, that we are freed from either dogmatism or capitulation to scepticism.'[52] MacIntyre here made a very important argument: rational and objective knowledge is knowledge about events in context. It is just this kind of knowledge which historians have experience and discipline – but no monopoly – in creating.[53]

In the sciences of being human, historical writing is the activity of the knowing subject knowing itself in context. By contrast, writing the history of the human sciences as a history of progress towards biological knowledge about human nature systematically elides the reflexive condition. It is writing not able to understand its own conditions of knowledge.

51 MacIntyre, 'Epistemological crises', p. 467.
52 *Ibid.*, p. 471.
53 Historians of science are thus attracted to 'historical epistemology' and 'historical ontology'. See Daston (ed.), *Biographies of Scientific Objects*; Hacking, *Historical Ontology*.

For the knowing subject to know itself requires transgression – 'going outside' any one way of thought. And this necessarily locates the place of thought in history. 'The moment philosophy arrogates to itself the right to question . . . the foundations of institutions and traditions; the moment that poetic language ceases to be confined by the rules of a well-defined game, or ceases to be the exorcism of transgression and becomes transgression itself; then culture acquires a historical dimension.'[54] We cannot discard history: it is the form of knowledge that self-understanding takes.

The form of knowledge characteristic of, but not exclusively possessed by, the human sciences is appropriate for the expression of meaning, significance and value, and it involves reflexive engagement with what such expression itself is. In Gadamer's idiom: 'Our starting point is that verbally constituted experience of the world expresses not what is present-at-hand, that which is calculated or measured, but what exists, what man recognizes as existent and significant. The process of understanding practiced in the moral sciences can recognize itself in this – and not in the methodological ideal of rational construction that dominates modern mathematically based natural science.'[55] As a consequence, part of what the human sciences analyse is the meaning, significance and value of the natural sciences. In so far as natural scientists themselves do this, as of course they sometimes do, they too contribute to the human sciences. This is clear, for example, in the case of Polanyi, who wrote about the kind of cultural, or spiritual, tradition which he thought necessary for the flourishing of the individual scientific mind.[56]

Kant, in his three critiques, provided what has become the philosophical reference point for discussion about different forms of knowledge. As Cassirer wrote, developing the Kantian tradition in a discussion of different symbol systems, 'It is characteristic of the nature of man that he is not limited to one specific and single approach to reality but can choose his

54 Starobinski, 'The critical relation', p. 119.
55 Gadamer, *Truth and Method*, p. 456.
56 Polanyi, *Personal Knowledge*.

point of view and so pass from one aspect of things to another.'[57] Weber expressed a similar argument as a sociological theory of interests, and Habermas returned to this in his philosophical and political critique of the positivist theory of knowledge. Positing different forms of knowledge was part of the ordinary language analysis of the distinction between explaining physical causes and understanding intentions as reasons for human action. It also entered debate among philosophers of science about reduction, whether the aim of scientific explanation is an account of all events in terms of matter and motion. Those who opposed reduction argued that there are different levels of explanation, each level appropriate for a particular level of complexity.[58] Thus, for example, social analysis is concerned with institutions (not individual people), natural selection theory with populations (not individual genes) and perception with the eye–brain system (not individual nervous impulses). A particular field may be reductionist but, as many people accept, science as a whole cannot be. What is significant to researchers varies, just as what is significant in ordinary life varies, and hence levels of explanation and kinds of knowledge likewise vary.

The collaborative writing of Arbib and Hesse, the one a scientist, the other a philosopher, illustrates the argument. The authors distinguished scientific research practice, generally committed to reductionism and the explanation of mental events by brain events, from a philosophical claim about what is real. They also recognised the active role scientists have in imposing a conceptual framework on what they study. 'Though we say cognitive science is reductionist, we in fact advocate a permissiveness with respect to ontology: there are all manner of levels of reality. What real things there are can depend on human constructions imposed on some "external stuff. "' They then introduced the 'schema' as a concept at the level of analysis appropriate for cognitive activity. Reference to the schema, they argued, allows the cognitive scientist to develop a different level of explanation than the brain scientist. 'We have many

57 Cassirer, *Essay on Man*, p. 170.
58 See Beckner, *Biological Way of Thought*.

different levels of description, including neural, mental, and social, and we find ways of illuminating any particular level of discourse by placing it within a higher level context and by seeking lower level mechanisms.'[59] Which level of description and analysis matters to a scientist depends on the purpose in hand.

There is in principle no contradiction in representing a phenomenon as at one and the same time a neural, cognitive, social and, indeed, ethical process. If there appear to be incompatibilities, these are the result of power struggles between disciplines not insuperable philosophical differences. What people know is inherently tied to the purposes they have as social beings. As Dewey argued: 'If we see that knowing is not the act of an outside spectator but of a participator inside the natural and social scene, then the true object of knowledge resides in the consequences of directed action . . . there will be as many kinds of known objects as there are kinds of effectively conducted operations of inquiry which result in the consequences intended.'[60]

The earlier example of narrative about crime and insanity made the same point. One form of knowledge, attributing the causes of violence to a disordered body, served one purpose, and another form, attributing crime to the wickedness of the soul, served another. The difficulty for the court, as so often in more mundane situations in daily life, was that the decision-making process required only one form of knowledge to apply, so to speak, in a particular social space. Legal rhetoric pictured empirical knowledge determining the decision; but in practice social purposes determined what kind of knowledge counted as empirical evidence. The ordinary person's sense of what is real, backed by the authority of natural science as empirical knowledge of what is real, and the need for cut and dried social decisions like reaching a legal verdict, all supported the habit of thinking that only one description ought

---

59 Arbib and Hesse, *Construction of Reality*, pp. 15, 65. The concept of 'the schema' itself has a history, associated with the psychologist Frederic Bartlett and the neurologist Henry Head.

60 Dewey, *Quest for Certainty*, p. 188.

to be true. But several stories were possible. This should occasion little surprise, since in other contexts we are perfectly at home with different descriptions of the same phenomenon: we feel little strain in living with the astronomer's and the poet's descriptions of the stars at night.

The simple truism is that life is many-dimensional and that descriptions of the world follow suit. It is possible to imagine, just about, a form of life with only one kind of purpose (say, brute survival), and hence one kind of knowledge; and, conversely, we can imagine, like George Orwell in *Nineteen Eighty-Four*, the unquestionable authority of one kind of knowledge fostering a one-dimensional way of life. What we thus imagine, however, is knowledge of, and a way of life for, something most unlike a person, as most people think of a person. There could be one form of knowledge only in a one-dimensional society, sustained by a being quite unlike what we call a human being. The psychologist B. F. Skinner, in his argument in *Beyond Freedom and Dignity* (1971), moved towards this kind of position (though, it should be remembered, with an avowed humanistic purpose). In utter contrast, there is a sensibility like Goethe's: 'No limit, no definition, may restrict the range or depth of the human spirit's passage into its own secrets or the world's.'[61]

In a one-dimensional way of life it would certainly not make sense to talk of purposes or morals of any kind. Life would simply be what it is. The world which we know is quite otherwise, structured as it is by what Dilthey called 'significance'. We expect explanation to address this. 'Explanation is interest-relative and context-sensitive. We expect an explanation of a fact to cite the factors that are *important* (where our notion of importance depends on the reason for asking the Why-question).'[62] Expressed in ordinary language, as the discussion of narrative emphasised, we might say that, given a purpose, we tell one story rather than another. Or we might say that all knowledge is part of a practical engagement which presupposes

---

61  Goethe, 'A friendly greeting' (1820), in *Scientific Studies*, p. 37.
62  Putnam, 'Beyond historicism', p. 297.

something about both the world and ourselves.[63] This leads back into the question of facts and values.

## The moral purposes of historical knowledge

Moral issues are not separable from cognitive ones. The act of knowing establishes a relationship which presupposes that the object of knowledge exists in a certain way to us. How we constitute this relationship, perhaps as a relationship centred on truth-telling, is a moral as well as cognitive matter. The point holds equally for historical and natural science knowledge.

Historians share the rhetoric and moral value of truth-telling with natural and social scientists. Indeed, the modern discipline has a distinguished record of exposing the narrow interest, or prejudice, which promotes one narrative about the past at the expense of others, which is conceptually incoherent or which ignores the full range of evidence. The training and collective empiricist ethos of the history discipline houses a potential for political critique through the use of empirical resources themselves. This is evident, for example, in contrasting histories of national identity. I earlier cited historical studies of the idea of race; any attempt now to use 'race' as an explanation of nationhood would be highly vulnerable to historical critique. This, it is worth noting, shows a kind of progress in the moral dimension of knowing.

The question of the moral content of historical knowledge has a very obvious place in the memory of terrible human actions, and there is a literature on precisely this matter in discussion of the Holocaust. If ever we were to learn from history, it would seem, it is by accurate memory of such an event. But memory is no neutral matter. For example, the actual process of cultivating memory of the Holocaust, some writers have suggested, renders it 'familiar', whereas the moral lesson which can be drawn from it may depend on its 'otherness' at the boundaries of human possibility. *How* we

---

63 On 'the ecology of knowledge', see Ben-Chaim, 'The disenchanted world and beyond'; Still and Good, 'Ontology of mutualism'.

remember 'changes the inquirer's relationship' to the object; and, in the case of the Holocaust, this determines one horizon of collective moral understanding.[64]

If historical work, the collective act of memory, changes our relationship to both the object of study and ourselves, this does not exclude objectivity. Obviously enough, like any other empirically minded scientist, the historian seeks the kind of objectivity which will persuade colleagues that a piece of writing gives the best possible account of events. At the same time, history-writing involves one human taking up a relationship to another, one act of writing taking up a relationship with another. It is the lesson of hermeneutics that this relationship always involves an act of interpretation, which is *for something*, and the shaping of this purpose, the 'lesson' for readers, give concrete life to the historian's 'encounter' (Heidegger's term) with the subject matter. Historical writing 'is itself a social practice which establishes a well-determined place for readers by redistributing the space of symbolic references and by thus impressing a "lesson" upon them; it is didactic and magisterial'.[65] It is like literature or psychoanalysis in this regard. Historians are the medium of dialogue between readers and the records in the present which they 'write up' as the past.

The writer, the sources and the reader engage in a kind of conversation. In historical writing about thought on being human, this conversation takes the form of an examination of what it is to be human. The historian-writer places the reader's own stance, her or his own cognitive and moral position, in relation with the reflexive activity that made that stance possible in the first place. As Habermas argued, writing about what he called the practical interest of the cultural (or human) sciences: they aim 'not at the comprehension of an objectified reality but at the maintenance of the intersubjectivity of mutual understanding, within whose horizon reality can first appear as something'.[66]

64 Pleasants, 'Concept of learning', p. 188.
65 Certeau, 'The historiographical operation', p. 87.
66 Habermas, *Knowledge and Human Interests*, p. 176.

History-writing, therefore, is a critical activity in (at least) three kinds of ways.[67] It can regulate the empirical authority of stories told. As the powerful response to those who would in some sense deny the Holocaust has shown, this can be impressive. Secondly, it can address the premises of thought about whole fields, shifting the way historians think about progress in knowledge. This occurred in history and the human sciences in response to modern feminism in the 1970s and 1980s, bringing the history of women and gender into the mainstream, and in the light of political goals like justice certainly making history into a field exhibiting progress. Lastly, history-writing can turn, reflexively, on itself. In doing this, it examines the premises of one way of understanding, and the form of life of which that understanding is an expression, in order to keep open the possibility of a different, and possibly better, form of understanding and way of life. 'It is in accomplishing self-reflection that reason grasps itself as interested.'[68] By contrast, most disciplinary knowledge is unreflexive about its most basic presuppositions. As Barry Sandywell, a philosopher of social science, observed: the '"sciences" and "disciplines" created within the representational mind-set may well be *reflective* but in their truncated ontological and epistemological self-awareness they sponsor a deeply *unreflexive* view of the world'.[69]

Finding out how 'reality can first appear as something' (to return to Habermas's phrase) is not usually among natural or, for the most part, social scientists' purposes. Working within a discipline, scientists assume that there is more than enough of interest without going over the ground by which people have come to their manner of knowing in the first place. And they are right for the purposes of disciplined knowledge in

67 Guy Oakes, 'Introduction' to Windelband, 'History and natural science', p. 167.
68 Habermas, *Knowledge and Human Interests*, p. 212 (italics removed). See, e.g., the historical work of the psychologist Jill G. Morawski: 'Self-regard and other-regard'; *Practicing Feminisms*; and 'Reflexivity and the psychologist'.
69 Sandywell, *Reflexivity and the Crisis of Western Reason*, p. 44.

their own fields. They are wrong, however, judged by other purposes. These other purposes, both cognitive and moral, come to the surface and preoccupy the human sciences, since in these sciences what it is to be human is itself the subject of knowledge. In part, the human sciences inherit religious and humanist traditions which posit being human as an end in itself. In part, these sciences share in the more recent forms of cultural life ('postmodernity') which expose terms like 'human', 'moral' and 'in itself' to the infinite regress of reflexive analysis. In both aspects, however, and especially in productive interaction between tradition and innovation, the history of the human sciences creates a dialogue with a questioning character which disciplinary knowledge does not have. This has a moral dimension. As Collingwood and Winch, as well as philosophers in the hermeneutic tradition have discussed, understanding other ways of life is necessary, logically, to extending and changing our own. This makes understanding itself a form of 'moral agency'.

The difference between disciplinary knowledge and explicitly moral knowledge can be brought out in an example, A. I. Tauber's discussion of what he called the 'moral agency' of knowing in the writings of Henry David Thoreau. Thoreau searched for a way of life, and the language to describe it, in which the very business of seeing nature would be a moral act. In the poetic language of his relationship with nature at Walden Pond and on the nearby Massachusetts rivers, he aspired to 'hear the dreaming of the frogs in a summer evening'.[70] His scientific contemporaries, however, though Thoreau wished it otherwise, judged that his descriptions of nature did not contribute much to knowledge as they understood it. The scientists valued a collective morality of rigour, objectivity and truth, embedded in rules for using language, over personal poetics and spiritual ambition. Discipline formation institutionalised the scientists' standards and excluded, at least from the scientific arena, the sort of moral knowledge which Thoreau created in writing about nature. All the same, the scientific

70 Quoted from Thoreau's journal (for 1851), in Tauber, *Thoreau and the Moral Agency of Knowing*, p. 176.

community, while excluding the poetics of individual moral agency from the public presentation of knowledge, fostered a collective moral agency in connection with truth-telling.

Where Thoreau did succeed was in inspiring himself, and countless individual readers, to experience knowing nature as a moral act. He carried on a dialogue, or 'conversation', between his own moral and sensuous being and the being that he found, saw or created in nature. His writing placed his being in nature and nature in himself. In Heidegger's terms this was an 'encounter': 'Letting something be encountered is primarily circumspective; it is not just sensing something, or staring at it. It implies circumspective concern, and has the character of being affected in some way.'[71] Historical and other scientific writing can make possible the same kind of 'circumspective concern', though, for the most part, authors are 'disciplined' and exclude such concern as part of the rhetoric of objectivity. Disciplinary writing still involves a stance. If, as someone exclusively committed to human biology might assert, there is need to replace all this reflexive talk by description of what exists in nature, that too exhibits a kind of moral agency. But I fear it is an agency which finds virtue in a monologue and in a one-dimensional form of life. It holds up a certain kind of knowledge of nature as the unique authority.

Knowledge has potential to change both the knowing person and what is known. The basic categories with which the sciences describe the human world – like person, society, mind, religion, law, economy, polity – originated historically in, and continue to give expression to, forms of life. There is no ground outside historically formulated knowledge on which to base eternal definitions. There is, therefore, both a cognitive and a moral obligation to be open to the possibility of difference. We cannot presume that a particular way of thinking adequately translates what is thought at another time or another place. Moreover, there is also plenty of empirical evidence from social anthropology and cultural history that it does not. In this connection, as Geertz said about ethnology, historical

71 Heidegger, *Being and Time*, p. 176.

work is 'enabling'.[72] If we do not imagine that something might be different, one way of life legislates the world. Ethnologists and historians, like philosophers and like scholars of the arts, need to keep open imagination for what is other. For the historian, 'the otherness of the other preserved in its difference and history can be, according to Paul Veyne's phrase, "the inventory of differences". Whence the dialectic between the alien and the familiar, the far and the near, at the very heart of the interest in communication.'[73] This 'dialectic' is crucial to knowing ourselves as well as others: 'we travel abroad to discover in distant lands something whose presence at home has become unrecognizable'.[74] When the historian examines the context of an individual or collective action, she engages in a dialogue with the actors as to what the subject is, and what is right and wrong, and she does so in dialogue also with herself. She takes a stance.

There is another link between ethnologists and historians: their interest in what is individual and particular. This interest has kept alive Windelband's distinction between nomothetic and idiographic knowledge. There are areas of learning, like those which concern the physician or biographer, and often enough the historian too, where the whole point of knowledge is to understand something individual. As Geertz sharply noted, in opposition to those scientists who think it most important to determine the universals of human nature, 'the notion that the essence of what it means to be human is most clearly revealed in those features of human culture that are universal rather than in those that are distinctive to this people or that is a prejudice we are not necessarily obliged to share'.[75]

Discernment of difference and knowledge of individual qualities is part of the foundation of moral action. Since fiction is a form richly endowed with descriptions of what is both other

---

72 Geertz, 'Uses of diversity'. For historical writing and recognition of 'the other', see Smith, 'History of human nature'.
73 Ricoeur, 'The narrative function', p. 295.
74 Certeau, *Practice of Everyday Life*, p. 50.
75 Geertz, 'Impact of the concept of culture on the concept of man', p. 43.

and particular, it is not surprising that much of the modern western moral imagination derives from it. Moral action requires a certain kind of orientation to the world, an orientation which, as one of its conditions, requires clarity of sight – not a set of rules but a vision. Shelley's statement of this is justly renowned: 'A man, to be greatly good, must imagine intensely and comprehensively; he must put himself in the place of another and of many others; the pains and pleasures of his species must become his own. The great instrument of moral good is the imagination.'[76] Similarly, Tauber, both physician and philosopher, noted, drawing on the work of Emmanuel Lévinas, that 'ethical responsibility to others rests on the recognition that in acting in the world, one inevitably changes it for others as well as for oneself . . . To *see* another then becomes an ethical act.'[77] History is not usually thought of in this connection. All the same, as the intimate link between 'history' and 'story' attests, to make something intelligible, to recognise its particularity and to view it ethically are not, finally, separable. In Gadamer's language, 'to recognize one's own in the alien, to become at home in it, is the basic movement of spirit'. The sciences of being human 'seek not to surpass but to understand the variety of experiences – whether of aesthetic, historical, religious, or political consciousness'.[78]

The blanket assertion of the superiority, let alone exclusive rights, of one form of knowledge rather than another makes no sense. As the account of narrative argued, knowledge is always for a purpose, and it is in relation to that purpose that we must judge whether it is adequate or not. 'The criteria that control "good talk" in science, in poetry, and in any other interpretive system depend on its point and purpose.'[79] In the light of this, I will comment on one important and influential argument to the effect that the natural sciences are decidedly ill-equipped, if not impotent, for 'good talk' about moral knowledge. It is the argument that the founders of modern science

---

76 Shelley, 'Defence of poetry', p. 517.
77 Tauber, *Confessions of a Medicine Man*, pp. 89–90.
78 Gadamer, *Truth and Method*, pp. 14, 99.
79 Arbib and Hesse, *Construction of Reality*, p. 181.

in the seventeenth century created a metaphysics which *excluded* moral subjects. This metaphysics has its best-known expression in Descartes's sharp separation of matter and mind and belief that the former but not the latter is a possible object of knowledge for natural science. One major outcome is the epistemological problem of how a mind, which is not matter, can know the material world.

The philosophers E. A. Burtt and Whitehead argued this thesis in the 1920s. They described the new astronomy and mechanics of the seventeenth century as a revolution in metaphysics – there was a new conception of what it is possible to have knowledge of, not just new knowledge. In Burtt's and Whitehead's view, whatever the gains there was also a cost, and the cost was that the new science excluded the knowing agent from the realm of nature. This split being human down the middle and left it incomprehensible how mind could be 'in' nature. As a consequence, when scientists later attempted to extend knowledge in the physical sciences to the human sphere they did not have the intellectual tools with which to carry it out. The result was either hugely inadequate mechanistic theories (like behaviourism) which conducted research as if people do not have minds, or the isolation of the different areas of scholarship concerned with matter and mind from each other, resulting in 'the two cultures'. Lastly, philosophers balked at understanding how two states, mind and matter, defined in mutually exclusive terms, could be said to interact.

As Joseph Needham once remarked, 'the problem of "mind and matter" has always been the skeleton in biology's cupboard'.[80] Sharing this view, the historian of science Robert M. Young turned to Darwin's work and the Victorian debate on 'man's place in nature'.[81] Here, he argued, when scientists made the major breakthrough in bringing knowledge of human beings into the orbit of knowledge of nature, the consequences of seventeenth-century metaphysics should be most apparent and the search for alternatives most productive. If Darwin so successfully brought the study of being human under the study

---

80 Needham, 'A biologist's view of Whitehead's philosophy', p. 201.
81 Young, *Mind, Brain, and Adaptation*; Young, *Darwin's Metaphor*.

of nature, how did he solve the problem of mind's place in nature? The Victorians themselves intensely debated the question. A century later, biologists and neuroscientists thought Darwin's work a decisive resolution. He had, they believed, shown, in the face of conservatism and prejudice, how it is indeed possible to understand human beings in the same mechanistic terms as the rest of nature. Here Young diverged, since it was his point, following Burtt and Whitehead, that it is these very mechanistic terms which cannot address the moral and knowing agency which matters at the centre of human life. The continuing presence of the mind–body question, along with the exclusion of moral categories from nature, suggested to him that only a new metaphysics would make possible knowledge of the human moral subject. Further, Young went on to put the Darwinian debates, Victorian and modern, in political context, consciously seeking a form of knowledge appropriate for humanistic purposes, and this sometimes led to conflict with modern Darwinists.

Earlier authors, heirs to a religious tradition, often stressed the contrast between nature and the true being of a person. This is emphatically not the present point. The older dualism of 'nature and man' is indeed untenable. Though many modern scientists therefore assume that human beings simply are objects in a biological story, it is equally reasonable to make historical knowledge the starting point for a biological story which is part of the world that human beings have created. If we take the latter road, however, we have to rethink knowledge in the human sciences, and perhaps also metaphysics. 'If we want to prevent the realm of humanity or history becoming a subcategory of Nature, we are going to have to admit to ourselves that Nature is in fact a subcategory of Humanity or History – that we are, after all, the authors of the system we call Nature . . . We are going to have to admit our own role in the constitution of reality, which in turn means admitting something quite fundamental about the nature of our knowing.'[82] This returns again to the reflexive theory of knowledge and to the challenge to describe people as a constitutive part

82 Evernden, *Social Creation of Nature*, p. 94.

of existence not as separate observers. This has been the message of philosophers otherwise as diverse as Heidegger, Wittgenstein, Whitehead and Rorty. If we seek knowledge which is not restricted to knowledge of changes without significance or meaning, which is all that the paradigm of mechanistic explanation set out in the seventeenth century can provide, we have to ground that knowledge in human action.

Vico's 'new science' is a kind of precedent for Burtt's and Whitehead's criticisms of the metaphysics of physical science. It was, it will be recalled, Vico's remarkable claim that certainty in knowledge of human actions is possible in a way that is not possible for physical events, of which God not man is the creator. Though the science of motion achieved precision in the seventeenth century, precision was not, Vico thought, to be confused with certainty and truth. This was also Burtt's and Whitehead's point: natural science has pursued precision at the expense of philosophical coherence, and it has done this by abstracting its subject matter from the totality of existence. A Victorian critic aptly noted that 'the essential characteristic of science is, that it submits to be partial for the sake of clearness'.[83] Whitehead therefore questioned the presumption that knowledge must begin from particular details: 'The whole notion of our massive experience conceived as a reaction to clearly envisaged details is fallacious. The relationship should be inverted. The details are a reaction to the totality.'[84] Burtt claimed that Descartes's incoherence on the mind–body relation made visible the consequences of abstracting from the world in order to achieve exactness in physical science.[85] Though expressing themselves in a different philosophical idiom, the phenomenologists, following Husserl, similarly argued for a return to knowledge grounded in the human 'life-world' not in abstractions.[86]

History, like the natural sciences, is noted for its focus on empirical particulars; but this does not gainsay the

83 Mozley, 'Philosophy, psychology, and metaphysics', p. 122.
84 Whitehead, *Modes of Thought*, pp. 148–9.
85 Burtt, *Metaphysical Foundations of Modern Physical Science*, pp. 308–24.
86 Husserl, 'The crisis of the sciences'.

fundamental importance of theoretical work. It is not only a matter of linking different fields together but of going outside fields of knowledge in order to understand what conditions of thought have made them possible and what kind of thought has thereby been excluded. 'Theoretical questioning . . . *does not forget*, cannot forget that in addition to the relationship of these scientific discourses to one another, there is also their common relation with what they have taken care to exclude from their field in order to constitute it . . . It is *the memory of this "remainder"*.'[87] 'Theoretical questioning' makes possible the reinsertion of specialised scientific knowledge into the knowledge-making process by which it became possible in the first place. In the modern natural sciences, this process separated the knowing agent from the object of knowledge, the sensuous richness of the perceptual world (the secondary qualities) from matter and motion in the physical world (the primary qualities), mind from matter and purposeful action from causal processes. The result is 'the celebration of a grand abstraction . . . torn free of all "irrelevant elements" – including human attributes such as color and smell, as well as meaning and purpose'.[88]

Taken too literally, the restriction of knowledge to the precise representation of physical causes shoves everything else into a bag labelled 'human'. Because all the things in this bag, all the subjective sensitivities and intuitions, are not fit objects of knowledge, they are left to the winds of fate, or, rather, to not very tender political and commercial depredation. 'The Cartesian split between the mental and physical life of individuals has become a split in Western culture between scientific "objectification" of a meaningless external world and a subjective and largely individualized world of meaning, sensibility, value, and action, to which the concepts of reality and truth have become almost inapplicable.'[89] This is the distinctive moral crisis of modernity, which so agitated Weber's generation, and which, if it agitated later intellectuals a little less,

87 Certeau, *Practice of Everyday Life*, p. 61.
88 Evernden, *Social Creation of Nature*, p. 49.
89 Arbib and Hesse, *Construction of Reality*, p. 160.

did so because of resignation not resolution. For Weber, to be modern is to be caught on the horns of a dilemma: to choose to objectify and lose meaning, or to assert feeling and lose reason.

A number of philosophers, including Burtt, Whitehead, Husserl and Heidegger, suggested that the only answer could be a new beginning in philosophy, starting from an account of being not burdened by abstraction. Other philosophers, like Foucault, however, questioned the possibility of a new metaphysics. By the early twenty-first century, prospects for a new metaphysics indeed looked remote.

I have tried to take the arguments in a different direction, away from philosophy and towards history. It is not necessary to choose between longing for a new metaphysics (and per-haps philosophical anthropology) and self-limiting irony. Even the most abstract work, whether in mathematical physics or in poetics, builds on what has gone before and presupposes a history. Even people who do not think of themselves as histor-ians in any sense nevertheless take a certain historical story about their own action and world for granted, and in this way, unconsciously if not consciously, they give themselves a certain historical nature. In self-consciousness about this his-tory, people converse with the content of what they themselves are experiencing and doing. Thus engaged, they change them-selves. In the case of an engagement with the history of what people think it is to be human, this conversation becomes explicit, 'out in the open' and itself the subject matter of sci-ence. And – the main point – historical knowledge of human reflexive nature does not start with abstractions in the same way as knowledge in the physical sciences. There is, as Vico argued, a kind of certainty about knowledge in the human sciences that the physical sciences do not possess. Historical knowledge of human reflexive activity, however incomplete, imprecise and open to reinterpretation, may claim to be con-crete and not abstract, and, in this respect, to be more exact than knowledge in the exact sciences.

This is not a matter of comparing different forms of know-ledge on some supposed scale of perfection. No one will question the unparalleled precision of the natural sciences. But knowledge about what makes a person significant, or an institution just, or a claim to truth persuasive, or a moment of perception beautiful, has a different character. It requires knowledge of particulars set in a story. Historically deracinated abstract knowledge, exemplified in the physical sciences, establishes no meaning, differentiates no shades of significance and points in no direction relevant to knowing what to do. It is knowledge about particulars, the place of people and events in a story, which opens such possibilities. 'What is abandoned [by the physical sciences] is ... to comprehend the given as such; [and what is needed is] not merely to determine the abstract spatio-temporal relations of the facts which allow them just to be grasped, but on the contrary to conceive them as the superficies, as mediated conceptual moments which come to fulfillment only in the development of their social, historical, and human significance.'[90] This, indeed, popularisers of natural science have tacitly recognised, since they have turned human evolution into a history of the supposed traits of human nature, like selfishness, and in the story given the traits meaning, significance and value. But the meaning, significance and value do not derive from the scientific knowledge; they derive from the historical culture which makes such stories look worth telling. To claim that some general thing called human nature exists outside and independently of the stories which are told about it is to turn an abstraction into a reality. 'Man' is an abstraction and does not exist. 'Man is to be defined neither by his innate capacities alone ... nor by his actual behaviors alone ... but rather by the link between them, by the way the first is transformed into the second, his generic potentialities focused into his specific performances. It is in man's *career*, in its characteristic course, that we can discern, however dimly, his nature.'[91] In so far as the evolutionary story by itself is supposed to tell us what human nature really is, it fails, because the story is abstract and detaches itself from the historical story about what has made the particulars of what people are. Whatever human inheritance is universally shared, it has no expression outside individual development. This is a

---

90 Adorno and Horkheimer, *Dialectic of Enlightenment*, pp. 26–7.
91 Geertz, 'Impact of the concept of culture on the concept of man', p. 52.

cognitive matter; but it is also a moral one, since to value a person is to value a particular person not the abstract entity, humankind. And we value a person, including ourselves, by placing her or him within a significant story which is about that person's life and surroundings, not by identifying the person as an evolved animal undifferentiated from others who share the same remote ancestors.

In this regard, the German proponents of *Geisteswissenschaft*, like Rickert, were surely correct. 'The *cultural importance* of an object . . . depends, as far as it is considered an integral *whole*, not on what it has in *common* with other real entities, but precisely on what *distinguishes* it from all the others.'[92] Here historical narrative achieves its purpose as the moral language of a stance towards what has individuality. Even those who disparage it implicitly recognise this, since they turn to it when they wish to assert something meaningful about the world to an audience whose sights are on more than specialised disciplinary goals. The discipline of history is an attempt to put human activity into dialogue with a more precise and examined account of what is going on. Historical work is thus, so to speak, the activity of reflection writ large.

92 Rickert, *Science and History*, p. 81. See Evernden, *Social Creation of Nature*, p. 117.

# Epilogue:
# on human self-creation

The history of thought about human nature leads to reflection on our most basic notions of what it is to be human. It is not possible to state what human nature is and be done with it; not least, the words 'human' and 'nature' have a history, and any empirical claim has meaning in the light of that history. It is not straightforward to say what the human sciences are, what they study and why their history should matter. This is because the field is, in principle, a field seeking to describe its own subject matter. The sciences of being human are open-ended: to utter a description, let alone to claim a truth, poten-tially re-creates the subject. 'Self-reflection is at the same time self-formation, self-creation.'[1] Past, present and future debate about the subject – man or woman, being human or human nature, culture or nature – gives the field its subject matter. Our being, as well as our understanding of it, is historical. In Gadamer's lapidary expression: 'In fact history does not be-long to us; we belong to it.'[2] We seek in history know-ledge of this 'belonging'. Historical narratives are basic to iden-tity and to human self-knowledge, collectively and individually.

Such argument may strike some ears as lacking harmony in contrast to the sweet melodies of an evolutionary past or the god who has laid down a well-defined and enduring human nature. For composers of evolutionary or religious songs, his-tory is a curious and perhaps to some people captivating set of variations on what human nature has done, but it is not

1 Groethuysen, 'Towards an anthropological philosophy', p. 87.
2 Gadamer, *Truth and Method*, p. 276.

the tune. I have argued, on the contrary, that evolutionary and religious ways of characterising human nature, along with any other ways we can think of, are themselves historically situated ways of being human. Constituted as reflexive action, consciousness cannot spring back into innocence; it must know itself historically or not know itself at all.

What, however, does this argument lead us to say about the positive content of being human? After all, it is central to the appeal of evolutionary, religious or psychoanalytic theories of human nature that they assign to this nature a clear and definite content, like aggression, sin or desire. The answer, in a troubled simplified word, is 'self-creation'. In Vico's expression, 'by its nature, the human mind is indeterminate'.[3] At first glance, the notion of self-creation is strikingly paradoxical. It brings to mind the fantastic tales of Baron Munchausen, who, amongst other things, claimed to have dug himself out of a deep hole by first going to fetch a spade. But something like this, I suggest, is the best metaphor we have for the human world and its history. As a result, history is, as Collingwood wrote, '"for" human self-knowledge ... The value of history, then, is that it teaches us what man has done and thus what man is.'[4]

Yet many natural and social scientists, as well as people who like to describe themselves as practical, have little patience with history. The action, they think, lies in investigating and changing the world in front of us, not with the record of what other people have thought, frequently enough mistakenly. If other mirrors distorted and failed, their mirror, they believe, will reflect true. But our knowledge is not aptly described by metaphors of reflection; it is more in the nature of a dialogue. Whatever view we take about the world to which our language refers, what we know about that world is always linguistically mediated. Understanding 'ties the interpreter to the role of a partner in dialogue'.[5] The appropriate image of human knowledge is of a conversation, not of a mirror. Neither self-reflection

3 Vico, *New Science*, p. 75 (§ 120).
4 Collingwood, *Idea of History*, p. 10.
5 Habermas, *Knowledge and Human Interests*, pp. 179–80.

nor intuition is enough for self-knowledge; rather, we must locate what we think we are in linguistic and semantic, as well as material, historical context. As Marx noted: 'Since . . . [the human being] comes into the world neither with a looking glass in his hand, nor as a Fichtian philosopher, to whom "I am I" is sufficient, man first sees and recognises himself in other men. Peter only establishes his own identity as a man by first comparing himself with Paul as being of like kind.'[6] Stories told about who or what a person is, or people are, reflect the conditions and purposes of their construction. 'Only in interaction within the world do . . . [people] create identities, and only in this creation of identities does the world of human beings take shape.'[7] If we accept this, then self-knowledge requires historical work, and no amount of biological knowledge, religious wisdom or psychoanalytic talk can substitute for it. Neither materialist natural science nor philosophical anthropology, the one dependent on an empiricist and the other on an idealist theory of knowledge, will by itself prove sufficient as an approach to human self-understanding. In addition, we require knowledge of how these approaches are themselves historically formed fields in the human sciences.

This was the starting point for Foucault's work in the 1960s: 'If one reduced man to his empirical side one could not account for the possibility of knowledge, and if one exclusively emphasized the transcendental one could not claim scientific objectivity nor account for the obscurity and contingency of man's empirical nature.'[8] It is this situation that has led to a reconsideration of earlier scholars like Cassirer and, more distantly, Kant. From their perspective, being human is activity giving 'form' to experience. Different kinds of activity – different purposes – shape experience in different ways; no one science possesses the truth. 'Cognition, language, myth and art: none of them is a mere mirror, simply reflecting images of inward or outward data; they are not indifferent media, but rather the true sources of light, the prerequisite of vision, and the

6 Marx, *Capital*, p. 59 note 1.
7 Nelson, Megill and McCloskey, 'Rhetoric of inquiry', pp. 7–8.
8 Dreyfus and Rabinow, *Foucault*, pp. 41–2.

wellsprings of all formation.'[9] History is knowledge of 'formation' in its concrete particularities; history of the human
sciences is knowledge of 'formation' as representation of the
human subject, that is, 'formation' as self-creation.

The metaphor of history as 'conversation' also builds on the
principle that 'every preposition has presuppositions that it
does not express'. This leads to the veiw, as Gadamer went on
to argue, that knowledge is finally about questions. The criterion of what is true relates to the question which is asked in
the first place, to the purposes of the person seeking knowledge.[10] And since every question is also an answer to something
that has come before, knowledge has the form of the dialectic, the recurring conversation taken by human life.

To be sure, much more needs to be said about self-creation.
This is hardly a simple matter, but we can feel some confidence
because at least since Vico, through Herder, Fichte and Marx,
down to humanistic voices of the twentieth century like Dewey
and Berlin, there has been rich investigation of the sense in
which we can say that human beings have, in freedom, created what they have in fact become. In Vico's re-creation of
Roman mythology, quoted earlier, the dramatic foundation of
the human world occurred when Jupiter thundered, 'waking'
in the human breast 'the moral effort or *conatus*' proper to a
free mind.[11]

As Vico understood, the notion of self-creation and the
notion of freedom are intimately connected. This is also evident
in modern writing; for example, Iris Murdoch's 'Our freedom
is not just a freedom to choose and act differently, it is also a
freedom to think and believe differently, to see the world differently, to see different configurations and describe them in
different words.'[12] We have somehow to assert such freedom
without getting bogged down in notorious problems. (I have
no intention of entering the labyrinthine world of philosophers'

9 Cassirer, *Philosophy of Symbolic Forms*, vol. 1, p. 93.
10 Gadamer, 'What is truth?', p. 42.
11 Vico, *New Science*, p. 311 (§ 689).
12 Murdoch, 'Metaphysics and ethics', pp. 72–3. Also Hampshire,
   *Thought and Action*, p. 177.

discussions of freedom and determinism, though what they call a compatibilist position, the belief that ordinary judgement about our individual or collective freedom to act is compatible with determinism in science, is here implicit.) Briefly, what sort of conception of freedom for self-creation appears to be required by the theory of historical knowledge in the human sciences which I have been putting forward?

In its historical origins, freedom is a political concept, but, at least since Vico and Kant, philosophers have also elaborated it in metaphysical and ethical enquiry, and it is in this context that I employ it. The sort of notion of freedom that appears here comes with the capacity to judge and act in the light of reflection on judgement and action, that is, with the meaning or 'significance' that judgement and action have. The very business of narrative and dialogue – in conversation, translation, the arts, the sciences or wherever – expresses 'the fact that we are *cultural beings*, endowed with the capacity and the will to take a deliberate attitude towards the world and to lend it *significance*'.[13] That freedom is circumscribed by 'finitude' no one doubts.[14] The result is, as Jean Starobinski wrote at the end of his study of the metaphor of action and reaction:

> The objective knowledge that claims to trace consciousness to its origins (biological, neurological), and that appears to dispossess it, must recognize itself as the product of a decision and as the bearer of future decisions: it cannot reverse the choice that produced it. This tentative freedom discovers that it was itself – through its interpretation of the world – that set out to discover its origins, to the point of hoping to see itself being born . . . We are the origin of our search for the origin. The circle closes, and another action begins.[15]

It is in this starting point in the phenomenology of consciousness where freedom, in the sense that is relevant here, lies. To whatever extent analysis and knowledge represents us as an effect, we remain also fundamentally a cause. They are

13  Weber, '"Objectivity" in social science', p. 81.
14  See Gadamer, 'Heritage of Hegel', pp. 50–2.
15  Starobinski, *Action and Reaction*, p. 371.

two poles of one circle – action and reaction, in Starobinski's metaphor.

Even the most methodologically rigorous social scientist, the most determinist biologist and the most ironical post-modernist still promote one way of viewing being human rather than another. They, too, act from positions within webs of 'significance'. And the commitment with which they act suggests that they, too, if tacitly, think they have some freedom to do so. Foucault, for example, both identified the horizons marking the limits of contemporary thought and lived as if they could be transgressed. He appeared to act out a life reflexively engaged with forming itself and the world even while describing the massive restraint on possibilities, for which the panopticon, or total institution, was a vivid symbol. In this regard, he joined a distinguished company of writers alert to the irony that people are committed to personal freedom of action while yet believing the world subject to chance and contingency. The revolutionary Russian writer Alexander Herzen, followed in this respect by Dostoevsky, conceded that much of the human world may be due to chance, but he still held that free will is a '"psychological or, if you wish, an anthropological reality," without which we could not function as social beings'.[16]

Freedom is also in some sense a consequence of language. Talk perpetuates and it transforms. 'To reach an understanding in a dialogue is not merely a matter of putting oneself forward and successfully asserting one's own point of view, but being transformed into a communion in which we do not remain what we were.'[17] Social scientist, biologist and post-modernist alike engage in determining significance, judging what enhances dialogue and deciding what an intervention is for. This activity, in which even self-declared determinists engage, is what I take freedom, in the sense that is relevant here, to be. By virtue of using language, people engage with a world of self-creation.

16 Kelly, *Toward Another Shore*, pp. 350–1, quoting Herzen. Also Sirotkina, 'A family discussion'.
17 Gadamer, *Truth and Method*, p. 379.

What appears to separate those who, like Dawkins or Pinker, want to find answers in biology to questions about what is human from their critics is a difference of belief about what gives meaning to what we say. The former appear to search for meaning in nature. This search, however anti-religious some of its modern expressions are, has its historical roots in religion, in the supposition that it makes sense to identify a *telos* beyond or prior to human existence, a meaning that nature gives us. For the critics, however, it is simply a category mistake to assert the meaning of anything except human activity, and human activity is historical. Meaning, by definition, they, and I, argue, comes into existence in human reflexive activity, mediated by language. If the science of biology is, indeed, 'life reflecting on itself', it is incomplete without knowledge of the nature of that reflection.[18] But that knowledge is knowledge of language, history, social life and what Susanne K. Langer called 'the essentially *transformational* nature of human understanding'.[19]

Langer's words, alluding to an essential human nature, recall the aspirations of philosophical anthropology. In the half-century since she wrote, philosophers called precisely these aspirations into question. Any claim about what is essential to being human is potentially a target for a demonstration that it is contingent on a particular 'regime of truth' (in language indebted to Foucault), or for deconstruction, showing that language refers to language and not to a world beyond language. It is foolish to pretend that these philosophies do not have force (though precisely what that force is, I have not debated). But it is also foolish to think that all that is left to do is to repeat what the masters of deconstruction have done. Recursive reflexivity is always possible; but, as a matter of living, we stop at a point appropriate for our purposes, be they the purposes of a scientific discipline or the purposes of a moral community.

In these circumstances, historical work in the human sciences has special importance. It is here that we can understand

18  Polanyi, *Personal Knowledge*, p. 347.
19  Langer, *Philosophy in a New Key*, p. xiv.

the meaning of claims about the human world, both as we find it and as we imagine it ideally might be. If, for example, as one philosopher of the social sciences stated, it is the 'values of human freedom and human integrity which ought to be the ultimate goal of the human sciences', we may make sense of this as an assertion about a desired way of life rather than as an assertion about the human essence.[20] We may also think that reflective history, rather than yet more speculative philosophical anthropology, will help us pursue ideal goals. As Dilthey might have argued, there is more access to self-knowledge through interpreting what people have actually done than through some supposed privileged insight into consciousness.

A contemporary (if you wish, postmodern) feeling for the contingency of things rubs uncomfortably against an ancient faith that wisdom lies with transcending the merely mortal for the eternal. There is still a deeply entrenched memory of Plato's metaphor of the cave, from which he hoped human-kind could return into the sun. Overcoming the antithesis of contingency and transcendence requires a rhetoric in which we give historical understanding of how the contrast has come about its due. There are precedents. Gadamer, for instance, asserted that 'history has a meaning in itself. What seems to speak against it – the transience of all that is earthly – is in fact its real basis. In impermanence itself lies the mystery of an inexhaustible productivity of historical life.'[21] If historical understanding points to the transience and perspectival character of human knowledge, it also points to the inexhaustible possibilities that still lie open. It is this variety, depth, mystery and open-ended possibility, realised in acts of reflection, which deterministic biological theories of human nature so conspicuously fail to capture.

What we are in need of is a language adequate to describe the self-forming of being human without tangling with an incredible idealism. This book argues that we have such a language, the language of history.

---

20 Manicas, *History and Philosophy of the Social Sciences*, p. 318.
21 Gadamer, *Truth and Method*, p. 202.

It is also relevant to note, however, that there is a literature in the natural sciences about self-forming or self-organisation, and there are scientists who would think that it is here that we should look to make objective sense of the notion of self-creation. The last quarter-century or so saw intense interest in self-perpetuating complex systems. (The basic physical exemplar is the laser beam, in which a photo-electric field generates a rhythmic, self-organising, self-perpetuating movement of atoms.) In the eyes of leading proponents, notably the physical chemist Il'ya Prigogine, the theory of self-organisation amounted to a new scientific revolution.[22] These scientists have formally specified the properties of systems that develop to a level of complexity in which they become self-organising and, in the process, intrinsic observers of their own capacity. Such self-observing activity is cognition. In this context, in 1972, the biologists Humberto Maturana and Francisco Varela referred to 'autopoiesis' (that is, self-production) and described the organism as 'an autopoietic machine [which] continuously generates and specifies its own organization through its operation as a system of production of its own components'.[23] An organism with a nervous system, in their view, has the capacity to generate reflexivity. This feature of life, in which the state of a system recursively alters that state, has, in the human organism, they argued, made possible the circle of linguistic interactions which continuously re-creates the world. Human linguistic reflexivity, or collective cognition, has the same structure as a living system. *'The logic of the* description *is the logic of the* describing *(living) system (and his cognitive domain)*.'[24] It follows that 'this circularity, this connection between action and experience, this inseparability between a particular way of being and how the world appears to us, tells us that *every act of knowing brings forth a world'*. They argued that 'we are constituted in language in a continuous becoming that we bring forth with others', and they

---

22 See Krohn, Küppers and Nowotny (eds), *Selforganization*. Niklas Luhmann elaborated a similar approach to social systems.
23 Maturana and Varela, 'Autopoiesis', p. 79.
24 Maturana, 'Biology of cognition', p. 39.

described the logic of this constitution as a property of biological systems.[25]

Whether or not such theories of self-organisation lead to profitable research programmes in the natural sciences is hardly for an outsider to judge. But it is possible to note, firstly, that the potential exists to examine the presuppositions of theories of biological autopoiesis, like any other theories, and to show that they are the outcome of cultural life. Secondly, it is open to question why we should privilege any one account of systems organisation, whether in physical, biological or social terms, over any other. Each account has its particular purposes; and there therefore may be reasons to prefer, in some contexts and for some purposes, a historical approach to human self-organisation over a biological or physical one.

Other authors have referred to '*anthropopoiésis*', i.e. human self-production, with acknowledgement to Herder rather than to the work of biologists. The classicist Claude Calame and cultural anthropologist Mondher Kilani hoped that the word would 'souligne bien cette idée de "faire", de "construire", de "fabriquer" des êtres humains, ou plus précisément des modèles d'êtres humains'.[26] This statement appears ambiguous. Is the underlying claim about the self-creation of being human or the possibility of reshaping notions of what it is to be human? Is the project to establish a philosophical anthropology focused on the self-creating being of being human (a project dependent on an unspecified metaphysics), or is it to argue for the cultural roots of all accounts of human life? What I have tried to show throughout this book, however, is that the ambiguity is not in fact real: it is precisely the capacity to change representations of what is human that makes possible changing what is human. This is the lesson of the analysis of reflexive knowledge. In this sense, history of the human sciences is indeed a science of *anthropopoiésis*, the study of the production of what is human.

It is possible to rephrase the notion of *anthropopoiésis* in Geertz's terms: 'We are, in sum, incomplete or unfinished

25 Maturana and Varela, *Tree of Knowledge*, pp. 26, 234–5.
26 Calame and Kilani, '*Anthropopoiésis*', p. 7. Also Blanckaert, 'La demande d'histoire', p. 19.

animals who complete or finish ourselves through culture
– and not through culture in general but through highly
particular forms of it.' Human life makes possible the establishment of meaning in 'particular forms' (whether 'Hopi and
Italian, upper-class and lower-class, academic and commercial'), and it is these 'particular forms' that have meaning by
virtue of their place in historical narratives.[27] It is not possible
in the modern western world to imagine a history of *anthropopoiésis* in general, in the manner in which Herder conceived of writing human history; unlike Herder's age, ours is
sceptical about establishing an agreed, let alone the true, ontology. Taylor, for example, wrote of his suspicion 'that no
satisfactory general formula can be found to characterize the
ubiquitous underlying nature of a self-interpreting animal'.[28]
This appears to me to be right. But such scepticism does not
at all detract from being able to imagine, as Herder also imagined, the work of weaving the fabric of particular cultures
open to reflection on themselves.

Modern discussions of human self-creation have a common
background in the philosophical idealism and romantic outlook which followed Kant and Herder. Philosophers turned
Kant's sober argument that human reason shapes what can
be known of the world into a metaphysical, and at times
romantic, quest for what can give rise to itself. Fichte's self-
positing 'I', Hegel's *'Geist'* (the *'in and for itself'*) and Marx's
'production', in their different ways, each shaped a notion of
self-forming. In its extreme expression, in Fichte's words, 'it
was the spontaneity of the human mind which brought forth,
not only the object of reflection . . . but also the form of reflection, the act of reflecting itself'.[29] There are great difficulties,
recognised even as Fichte wrote, in making sense of this. One
very influential route to an answer lay in interpreting what
Fichte called 'spontaneity' as itself a development, a historical
process, through which full reflection comes into being and

27  Geertz, 'Impact of the concept of culture on the concept of man',
    p. 49.
28  Taylor, *Sources of the Self*, p. 112.
29  Fichte, *Science and Knowledge*, p. 198.

makes possible the ideal of what is most fully human. Historically, part of this step involved comparing organic epigenesis
(the development of the embryo) and the reflective generation
of the 'I' out of itself.[30] In the romantic vision, the world as a
whole, living organisms, individual 'genius' (the character or
talent of a person) and the culture of a people – each had a
gestation. Each level of reality, it was thought, expresses
autopoiesis. In this argument, the idea of organic teleology, or
the idea that living things embody a purpose which is its own
ground and is not itself caused, played a prominent part. The
'desire' of the organism to be – to come into existence and to
maintain itself in existence – appeared to be a profound metaphor, or even literal basis, for the self-creative nature of the
human spirit. Subsequently, Hegel, followed in this regard
by Marx, recognised that, whatever the historical nature of
being, there is nevertheless a distinction to be made, since the
conscious reflection of human beings on themselves changes
the nature of this gestation. Human beings 'are never simply
what they "are," like other animals, but are as they take themselves to be, which, so Hegel thought he had shown, is
developmental in both the social and historical senses'.[31] The
history of human autopoiesis must include the history of reflection on what humans think themselves to be and not only
what, by gestation, they become. By expressing the relationship between human self-forming and the reflexive conditions
of knowledge, and by understanding this relationship as a
historical one, these writers shaped much of modern philosophy. Not least, they made reflexivity inescapable. For Hegel,
it is intrinsic to the notion of 'Spirit' that it gains 'knowledge
of itself as Spirit, i.e. it must be an *object* to itself'.[32] And
Marx, in the midst of his attack on Hegel's abstraction, wrote:
'And as everything natural must *come into being*, so man
also has his process of origin in *history*. But for him history
is a conscious process, and hence one which consciously

30 See Müller-Sievers, *Self-Generation*.
31 Pinkard, *German Philosophy*, p. 294.
32 Hegel, *Phenomenology of Spirit*, p. 14 (§ 25).

supersedes itself.'[33] The idea that normative authority is self-legislating, and that meaning and significance have come into being in the expressive activity of consciousness, are lasting legacies in modern thought.[34]

In the twentieth century, the idea of self-creation reached its limit in the philosophical existentialism of Jean-Paul Sartre. His statements were to the effect that free action ontologically precedes being translated into a highly influential practical ethic holding people responsible, in the most radical sense, for what they are: 'No limits to my freedom can be found except freedom itself or, if you prefer, that we are not free to cease being free.'[35] With a dogged consistency, it was asserted that any form of human science (psychoanalysis was a favourite target) is an act of bad faith, a culpable self-deception, since any science attributes action to causes rather than to its ground in existential freedom. Though the coherence of Sartre's own views was questioned, a radical idea of self-creation became common currency in the second half of the twentieth century through his work.[36] And it persisted, if toned down, in the literature of self-shaping.

Further to this, the idealist notion of desire, to which Hegel among others had drawn attention, was to have an influential life. Conceived in the romantic period as sharing a nature with the non-materially caused form of life itself, late nineteenth-century authors shaped the notion of desire into something more analogous to a material force or energy. Nietzsche re-created the notion as 'the will to power', and, incidentally, retained a considerable amount of the ambiguity about whether such terms referred to something ideal as opposed to something material. Freud built the notion into the base of psychoanalytic thought, with some of Nietzsche's ambiguity about its nature, but with clearer reference to inherited biological forces. It was in this sense that Cassirer, for example, understood desire, and he equated it with the

---

33 Marx, 'Economic and philosophical manuscripts', p. 391.
34 See Pinkard, *German Philosophy*, pp. 358–62.
35 Sartre, *Being and Nothingness*, p. 439.
36 See Izenberg, *Existentialist Critique of Freud*, pp. 266–74.

first constructive energies of the human mind. (To learn about these, he turned to Freud.)[37] Subsequently, 'desire', through the influence of Lacan, passed into the jargon of intellectual life, where it did service denoting the ungrounded being, which exists for itself and beyond which we cannot go in seeking explanation, of the purposive aspirations, fundamentally unconscious, of human action. Modern authors would hardly write, along with Herder, that there is 'the force of thinking, of acting according to an ideal of perfection, [which] is the essence of the soul'.[38] Nevertheless, they still strive for language adequate to represent the purpose-directed and significance-bearing form of conscious or unconscious human action, and in this regard the language of desire links the modern world with the romantic age.

Cassirer's reference to desire was at the heart of his attempt to reconcile a Kantian theory of knowledge – an idealist appreciation of the form-giving nature of perception – with modern biology. This attempt at a philosophical anthropology was not successful. But the task which Cassirer set himself was clear: to reconcile the teleology of human striving (that is, 'desire') with mechanistic biochemical explanation. For most modern scientists, this is a much too idealist formulation. They are much happier with the translation of the question into the problem of consciousness, that is, the problem of how to specify mental events (which show intentionality) as the outcome of brain processes, a problem about which history, in contrast to neuroscience and neurophilosophy, has nothing to say. But, it can be argued, whatever empirical progress there is in neuroscience, this still leaves the problem of stating how judgements of significance, which appear intrinsic to conscious reflection, relate to knowledge of material nature. This is an epistemological, and perhaps moral, question, for which history does indeed provide resources.

For Cassirer, it is precisely the forward-looking nature of consciousness that has given us a historical past. What we call the past is the story of people looking forward, and in this

---

37 Cassirer, *Philosophy of Symbolic Forms*, vol. 2, p. 157.
38 Herder, 'On cognition and sensation' (1775), p. 183.

story we confirm that we, too, look forward. 'For the meaning of historical time is built not solely from recollection of the past, but no less from anticipation of the future . . . Only a being who wills and acts, who reaches into the future and determines the future by his will, can have a "history"; only such a being can *know* of history because and insofar as he continuously produces it.'[39] This 'continuous production' is the autopoiesis to which other authors have referred. As human activity looks forward, it leaves behind what it subsequently knows as the past.

This sketch is an appreciation of the search for knowledge of being human as a historical process of self-forming. I am not a metaphysician who can conjure up knowledge of the ungrounded being of what is human – be it soul, desire, the linguistic sign, the act, the laws of matter or whatever. What there is to know about being human, in this argument, is the historical form the self-forming has actually taken. This leads into the history of the sciences. The scientific disciplines, whether biology, sociology or history, give stability, even the appearance of permanence, to assumptions about the world. They provide a base, itself taken for granted, on which to build up knowledge and make progress in practical ways of life. But this base is in principle always vulnerable to rational critique. It depends on socially embedded purposes whether or not there is such a critique and – the political question – by whom and for whom. Because a reflexive turn is always possible, and with it a new human self-interpretation or self-formation, knowledge in the human sciences must seek its ground in this reflexive activity not in some ground outside what is human. This ground in reflexive activity is, however, not fixed but itself an active process. The appropriate myth is an endless quest not a hunt for buried treasure. There is no 'right' or 'objective' or 'neutral' picture of human nature; each picture is part of the self-creative activity of being human.

Reflection, however, is not abstract. Sartre, for example, struggled at length, and by general agreement unsatisfactorily, to make what he said about existential freedom relate to

39 Cassirer, *Philosophy of Symbolic Forms*, vol. 3, p. 182.

what he knew of material and political circumstance. If as a
matter of fact we value a certain way of life, then we will tell
stories that make sense of what we value and criticise stories
that do not. Thus I have criticised the genre of scientific story-
telling which claims priority, or even exclusivity, for biological
theories of human nature as the basis of the human sciences.
From the viewpoint taken here, such stories are part of a
world-making in which people become biological objects.
But, were such a world to be created, a world excluding other
stories, there would be no possibility for scientists to promote
their stories on the grounds that they are objective and true.
Nor would there be any significance in anything. Stories about
people as biological objects cannot sustain the way of life in
which norms like objectivity and truth have authority and in
which things and events have significance. We need other
knowledge.

There are different forms of knowledge because we have
diverse purposes and ask diverse questions – 'for no theory or
picture is complete for *all* purposes' – not because there are
different objects in the world.[40] Historical knowledge has its
place in natural science, human science and everyday speech
alike. As Collingwood argued, 'History is no longer in any
special sense knowledge of the human as opposed to the natural world. It is simply that knowledge of facts or events as
they actually happen, in their concrete individuality.'[41] As a
consequence, it is possible to envisage history as a form of
knowledge bridging the institutional divisions between the
natural sciences, social sciences and humanities. We need
history in order to understand, in the fullest sense of 'to
understand', *any* of the forms that knowledge takes.

Shelley's extravagant claim, that 'poets are the unacknow-
ledged legislators of the World', has unexpected force.[42] But,
while Shelley himself thought of poets revealing to society
the law of the ideal, I have been discussing 'legislation' as

40 Putnam, *Reason, Truth and History*, p. 147.
41 Collingwood, *Idea of History*, p. 199.
42 Shelley, 'Defence of poetry', p. 535. See Abrams, *The Mirror and
the Lamp*, pp. 331–3.

a *process*, a historical coming-into-being. This, to use traditional words – though not to refer to an entity – is the life of
the spirit. The *poiesis*, or human production, of speaking or
writing turns being human into the subject of a story. In this
self-creation, people lay down the forms of knowledge that
make it possible to claim truth, goodness or beauty for some
aspect of their world.

To be human is to be party to the history of self-formation.
This activity manifestly includes many tasks, and each task
has its appropriate tools, the concepts that make possible one
kind of knowledge rather than another. These tasks include
renewing or finding new expression, including disciplined
expression in the human sciences, for what we think to be of
value – including the pursuit of science itself.

# Bibliography

Abrams, M. H., *The Mirror and the Lamp: Romantic Theory and the Critical Tradition* (Oxford: Oxford University Press, 1953).

Adorno, Theodor and Max Horkheimer, *Dialectic of Enlightenment* (1947), trans. John Cumming (London: Verso, 1979).

Anderson, Perry, 'Components of the national culture' (1968), in *English Questions* (London: Verso, 1992).

Anthony, Louise M. and Charlotte E. Witt (eds), *A Mind of One's Own: Feminist Essays on Reason and Objectivity* (1992), 2nd edn (Boulder: Westview Press, 2002).

Arbib, Michael A. and Mary B. Hesse, *The Construction of Reality* (Cambridge: Cambridge University Press, 1986).

Azouvi, François, 'Physique and moral', in John P. Wright and Paul Potter (eds), *Psyche and Soma: Physicians and Metaphysicians on the Mind–Body Problem from Antiquity to Enlightenment* (Oxford: Clarendon Press, 2000).

Baker, Keith Michael, *Condorcet: From Natural Philosophy to Social Mathematics* (Chicago: University of Chicago Press, 1975).

—— 'Enlightenment and the institution of society: notes for a conceptual history', in William Melching and Wyger Velema (eds), *Main Trends in Cultural History: Ten Essays* (Amsterdam: Rodopi, 1994).

Barham, Peter, *Schizophrenia and Human Value: Chronic Schizophrenia, Science and Society* (1984), new edn (London: Free Association Books, 1993).

Bazerman, Charles, 'Codifying the social scientific style: the APA Publication Manual as a behaviorist rhetoric', in John S. Nelson, Allan Megill and Donald N. McCloskey (eds), *The Rhetoric of the Human Sciences: Language and Argument in Scholarship and Public Affairs* (Madison: University of Wisconsin Press, 1987).

—— *Shaping Written Knowledge: The Genre and Activity of the Experimental Article in Science* (Madison: University of Wisconsin Press, 1988).

Beckner, Morton, *The Biological Way of Thought* (1959) (Berkeley: University of California Press, 1968).

Beer, Gillian, 'Problems of description in the language of discovery', in George Levine (ed.), *One Culture: Essays in Science and Literature* (Madison: University of Wisconsin Press, 1987).

Ben-Chaim, Michael, 'The disenchanted world and beyond: toward an ecological perspective on science', *History of the Human Sciences*, 11:1 (1998), 101–27.

Berlin, Isaiah, 'Does political theory still exist?', in Peter Laslett and W. G. Runciman (eds), *Philosophy, Politics and Society (Second Series)* (Oxford: Basil Blackwell, 1962).

—— 'Vico and Herder' (1976), in Henry Hardy (ed.), *Three Critics of the Enlightenment: Vico, Hamann, Herder* (Princeton: Princeton University Press, 2000).

—— *The First and the Last*, intro. Henry Hardy (London: Granta Books, 1999).

Bernal, J. D., *The World, the Flesh and the Devil: An Enquiry into the Future of the Three Enemies of the Rational Soul* (London: Kegan Paul, Trench, Trubner, 1929).

Bernstein, Richard J., *The Restructuring of Social and Political Theory* (1976) (London: Methuen, 1979).

Bijker, Wiebe E., Thomas P. Hughes and Trevor J. Pinch (eds), *The Social Construction of Technological Systems: New Directions in the Sociology and History of Technology* (Cambridge, Mass.: MIT Press, 1987).

Billig, Michael, 'Psychology, rhetoric and cognition', in Richard H. Roberts and James M. M. Good (eds), *The Recovery of Rhetoric: Persuasive Discourse and Disciplinarity in the Human Sciences* (London: Bristol Classical Press, Duckworth, 1993).

Blanckaert, Claude, 'La demande d'histoire: du détour au parcours', in Claude Blanckaert, Loïc Blondiaux, Laurent Loty, Marc Renneville and Nathalie Richard (eds), *L'Histoire des sciences de l'homme: trajectoire, enjeux et questions vives* (Paris: L'Harmattan, 1999).

Blondiaux, Loïc and Nathalie Richard, 'A quoi sert l'histoire des sciences de l'homme?', in Claude Blanckaert, Loïc Blondiaux, Laurent Loty, Marc Renneville and Nathalie Richard (eds), *L'Histoire des sciences de l'homme: trajectoire, enjeux et questions vives* (Paris: L'Harmattan, 1999).

Blumenberg, Hans, *The Legitimacy of the Modern Age* (1966), trans. Robert M. Wallace (Cambridge, Mass.: MIT Press, 1983).

Bouwsma, William J., 'From history of ideas to history of meaning', *Journal of Interdisciplinary History*, 12 (1981), 279–91.

Boyle, Nicholas, *Goethe: The Poet and the Age, vol. 2, Revolution and Renunciation (1790–1803)* (Oxford: Clarendon Press, 2000).

Braunstein, Jean-François, 'La critique Canguilhemienne de la psychologie', *Bulletin de psychologie*, 52:2 (1999), 181–90.

—— 'Bachelard, Canguilhem, Foucault: le "style français" en épistémologie', in Pierre Wagner (ed.), *Les Philosophes et la science* (Paris: Gallimard, 2002).

—— 'La philosophie des sciences d'Auguste Comte', in Pierre Wagner (ed.), *Les Philosophes et la science* (Paris: Gallimard, 2002).

Breuer, Josef and Sigmund Freud, *Studies on Hysteria* (1895), in *The Standard Edition of the Complete Psychological Works of Sigmund Freud*, vol. 2, ed. and trans. James and Alix Strachey (London: Hogarth Press, 1955).

Burnham, John C., 'Assessing historical research in the behavioral and social sciences: a symposium' [editor's introduction], *Journal of the History of the Behavioral Sciences*, 35 (1999), 225–6.

Burtt, Edwin Arthur, *The Metaphysical Foundations of Modern Physical Science: A Historical and Critical Essay* (1924), 2nd edn (London: Routledge & Kegan Paul, 1932).

Calame, Claude and Mondher Kilani, '*Anthropopoiésis*: introduction', in Claude Calame and Mondher Kilani (eds), *La Fabrication de l'humain dans les cultures et en anthropologie* (Lausanne: Payot, 1999).

Callon, Michel and Bruno Latour, 'Unscrewing the big Leviathan: how actors micro-structure reality and how sociologists help them to do so', in K. Knorr-Cetina and A. V. Cicourel (eds), *Advances in Social Theory and Methodology: Toward an Integration of Micro- and Macro-Sociologies* (Boston/London: Routledge & Kegan Paul, 1981).

Canguilhem, Georges, 'Qu'est-ce que la psychologie?' (lecture 1956), in *Etudes d'histoire et de philosophie des sciences*, 7th edn (Paris: J. Vrin, 1994).

Carr, David, *Time, Narrative, and History* (Bloomington: Indiana University Press, 1996).

Carrithers, David, 'The Enlightenment science of society', in Christopher Fox, Roy Porter and Robert Wokler (eds), *Inventing Human Science: Eighteenth-Century Domains* (Berkeley: University of California Press, 1995).

Carroy, Jacqueline, Nicole Edelman, Annick Ohayon and Nathalie Richard (eds), *Les Femmes dans les sciences de l'homme (XIXᵉ–XXᵉ siècles): inspiratrices, collaboratrices ou créatrices?* (Paris: Seli Arslan, 2005).

Cassirer, Ernst, *The Philosophy of Symbolic Forms* (1923–29), trans. Ralph Mannheim, 3 vols (New Haven: Yale University Press, 1955–57).

—— *The Logic of the Humanities* (1942), trans. Clarence Smith Howe (New Haven: Yale University Press, 1961).

—— *An Essay on Man: An Introduction to a Philosophy of Human Culture* (1944) (New Haven: Yale University Press, 1972).

Certeau, Michel de, 'Making history: Problems of method, problems of meaning' (1970), in *The Writing of History*, trans. Tom Conley (New York: Columbia University Press, 1988).

—— 'The historiographical operation' (1974), in *The Writing of History*, trans. Tom Conley (New York: Columbia University Press, 1988).

—— *The Practice of Everyday Life*, trans. Steven F. Rendell (Berkeley: University of California Press, 1984).

Churchland, Paul M., 'Eliminative materialism and the propositional attitudes' (1981), in *A Neurocomputational Perspective: The Nature of Mind and the Structure of Science* (Cambridge, Mass.: MIT Press, 1989).

—— *Matter and Consciousness: A Contemporary Introduction to the Philosophy of Mind* (Cambridge, Mass.: MIT Press, 1984).

Clark, William, 'The death of metaphysics in enlightened Prussia', in William Clark, Jan Golinski and Simon Schaffer (eds), *The Sciences in Enlightened Europe* (Chicago: University of Chicago Press, 1999).

Clifford, W. K., 'Body and mind', *Fortnightly Review*, new series 16 (1874), 714–36.

Coleman, Janet, *Ancient and Medieval Memories: Studies in the Reconstruction of the Past* (Cambridge: Cambridge University Press, 1992).

Collingwood, R. G., *The Idea of History* (1946) (Oxford: Oxford University Press, 1961).

Collini, Stefan, '"Discipline history" and "intellectual history". Reflections on the historiography of the social sciences in Britain and France', *Revue de synthèse*, 4th series 109:3–4 (1988), 387–99.

—— *Public Moralists: Political Thought and Intellectual Life in Britain 1850–1930* (Oxford: Clarendon Press, 1991).

Collini, Stefan, Donald Winch and John Burrow, *That Noble Science of Politics: A Study in Nineteenth-Century Intellectual History* (Cambridge: Cambridge University Press, 1983).

Comte, Auguste, *The Essential Comte: Selected from Cours de philosophie positive*, ed. S. Andreski, trans. Margaret Clarke (London: Croom Helm, 1974).

Cosmides, Leda, John Tooby and Jerome H. Barkow, 'Introduction: evolutionary psychology and conceptual integration', in Jerome H. Barkow, Leda Cosmides and John Tooby (eds), *The Adapted Mind: Evolutionary Psychology and the Generation of Culture* (New York: Oxford University Press, 1992).

Coulter, Jeff, *The Social Construction of Mind: Studies in Ethnomethodology and Linguistic Philosophy* (London: Macmillan, 1979).

—— *Mind in Action* (Cambridge: Polity Press, 1989).

Danto, Arthur C., *Analytical Philosophy of History* (Cambridge: Cambridge University Press, 1965).

Danziger, Kurt, *Constructing the Subject: Historical Origins of Psychological Research* (Cambridge: Cambridge University Press, 1990).

—— 'Concluding comments', in A. C. Brock, J. Louw and W. van Hoorn (eds), *Rediscovering the History of Psychology: Essays Inspired by the Work of Kurt Danziger* (New York: Kluwer/Plenum, 2004).

Darwin, Charles, *The Descent of Man, and Selection in Relation to Sex* (1871) (Princeton: Princeton University Press, 1981).

Daston, Lorraine (ed.), *Biographies of Scientific Objects* (Chicago: University of Chicago Press, 2000).

Dawkins, Richard, *The Selfish Gene* (1976), 2nd edn (Oxford: Oxford University Press, 1989).

Dehue, Trudy, *Changing the Rules: Psychology in The Netherlands, 1900–1985* (1990), trans. Michael O'Loughlin (Cambridge: Cambridge University Press, 1995).

Dewey, John, 'The need for social psychology', *Psychological Review*, 24 (1917), 266–77.

—— *The Quest for Certainty: A Study of the Relation of Knowledge and Action. Gifford Lectures 1929* (London: George Allen & Unwin, 1930).

Dilthey, Wilhelm, *Introduction to the Human Sciences: An Attempt to Lay a Foundation for the Study of Society and History* (1883), trans. and intro. Ramon J. Betanzos (London: Harvester Wheatsheaf, 1988).

—— 'Ideas concerning a descriptive and analytic psychology' (1894), trans. Richard M. Zaner, in *Descriptive Psychology and Historical Understanding*, intro. Rudolf A. Makkreel (The Hague: Martinus Nijhoff, 1977).

—— 'The construction of the historical world in the human studies' (1910), in *Selected Writings*, ed. and trans. H. P. Rickman (Cambridge: Cambridge University Press, 1976).

Dreyfus, Hubert L. and Paul Rabinow, *Michel Foucault: Beyond Structuralism and Hermeneutics* (Brighton: Harvester Press, 1982).

Dumit, Joseph, *Picturing Personhood: Brain Scans and Biomedical Identity* (Princeton: Princeton University Press, 2004).

Epstein, Mikhail N., *After the Future: The Paradoxes of Postmodernism and Contemporary Russian Culture*, trans. Anesa Miller-Pogacar (Amherst: University of Massachusetts Press, 1995).

Eribon, Daniel, *Michel Foucault* (1989), trans. Betsy Wing (Cambridge, Mass.: Harvard University Press, 1991).

Ermarth, Michael, *Wilhelm Dilthey: The Critique of Historical Reason* (Chicago: University of Chicago Press, 1978).

Evernden, Neil, *The Social Creation of Nature* (Baltimore: The Johns Hopkins University Press, 1992).

Farber, Paul Lawrence, *The Temptations of Evolutionary Ethics* (1994) (Berkeley: University of California Press, 1998).

Fichte, J. G., *Science and Knowledge: With the First and Second Introductions* (3rd edn 1802), ed. and trans. Peter Heath and John Lachs (Cambridge: Cambridge University Press, 1982).

Flanagan, Jr, Owen J., 'Psychology, progress, and the problem of reflexivity: a study in the epistemological foundations of psychology', *Journal of the History of the Behavioral Sciences*, 17 (1981), 375–86.

—— *Consciousness Reconsidered* (Cambridge, Mass.: MIT Press, 1992).

Foucault, Michel, *The Order of Things: An Archaeology of the Human Sciences* (1966), trans. Alan Sheridan (London: Tavistock, 1970).

—— 'Nietzsche, genealogy, history' (1971), trans. Donald F. Bouchard and Sherry Simon, in Paul Rabinow (ed.), *The Foucault Reader* (New York: Pantheon Books, 1984).

—— 'Truth and power' (1977), trans. Colin Gordon, in Colin Gordon (ed.), *Power/Knowledge: Selected Interviews and Other Writings 1972–1977* (Brighton: Harvester Wheatsheaf, 1980).

—— 'What is enlightenment?', trans. Catherine Porter, in Paul Rabinow (ed.), *The Foucault Reader* (New York: Pantheon Books, 1984).

Fox, Christopher, Roy Porter and Robert Wokler (eds), *Inventing Human Science: Eighteenth-Century Domains* (Berkeley: University of California Press, 1995).

Fraser, Mariam, 'The nature of Prozac', *History of the Human Sciences*, 14:3 (2001), 56–84.

Gadamer, Hans-Georg, 'Truth in the human sciences' (1954), trans. Brice R. Wachterhauser, in Brice R. Wachterhauser (ed.), *Hermeneutics and Truth* (Evanston: Northwestern University Press, 1994).

—— 'What is truth?' (1957), trans. Brice R. Wachterhauser, in Brice R. Wachterhauser (ed.), *Hermeneutics and Truth* (Evanston: Northwestern University Press, 1994).

—— *Truth and Method* (1960, 5th edn 1986), 2nd English edn, trans. revised Joel Weinsheimer and Donald G. Marshall (New York: Continuum, 1998).

—— 'The universality of the hermeneutical problem' (1966), in David E. Linge (ed. and trans.), *Philosophical Hermeneutics* (Berkeley: University of California Press, 1976).

—— 'Hermeneutics as practical philosophy' (1976), in *Reason in the Age of Science*, trans. Frederick G. Lawrence (Cambridge, Mass.: MIT Press, 1981).

—— 'The heritage of Hegel' (1979), in *Reason in the Age of Science*, trans. Frederick G. Lawrence (Cambridge, Mass.: MIT Press, 1981).

Gallie, W. B., *Philosophy and the Historical Understanding* (London: Chatto & Windus, 1964).

Geertz, Clifford, 'The growth of culture and the evolution of mind' (1962), in *The Interpretation of Cultures: Selected Essays* (New York: Basic Books, 2000).

—— 'The impact of the concept of culture on the concept of man' (1966), in *The Interpretation of Cultures: Selected Essays* (New York: Basic Books, 2000).

—— 'Thick description: Toward an interpretive theory of culture' (1973), in *The Interpretation of Cultures: Selected Essays* (New York: Basic Books, 2000).

—— 'The uses of diversity', in Sterling M. McMurrin (ed.), *The Tanner Lectures on Human Values*, vol. 7 (Salt Lake City: University of Utah Press, and Cambridge: Cambridge University Press, 1986).

—— *Works and Lives: The Anthropologist as Author* (Cambridge: Polity Press, 1988).

—— 'The strange estrangement: Taylor and the natural sciences', in James Tully (ed.), with assistance of Daniel M. Weinstock, *Philosophy in an Age of Pluralism: The Philosophy of Charles Taylor in Question* (Cambridge: Cambridge University Press, 1994).

Geras, Norman, *Marx and Human Nature: Refutation of a Legend* (London: Verso and NLB, 1983).

Gergen, Kenneth J. and Mary M. Gergen (eds), *Historical Social Psychology* (Hillsdale, N.J.: Lawrence Erlbaum Associates, 1984).

Giddens, Anthony, *Central Problems in Social Theory: Action, Structure and Contradiction in Social Analysis* (London: Macmillan, 1979).

—— *New Rules of Sociological Method: A Positive Critique of Interpretive Sociologies* (1976), 2nd edn (Cambridge: Polity Press, 1993).

Goethe, Johann Wolfgang von, *Scientific Studies*, ed. and trans. Douglas Miller (New York: Suhrkamp, 1988).

Good, James M. M., 'Quest for interdisciplinarity: the rhetorical constitution of social psychology', in Richard H. Roberts and James M. M. Good (eds), *The Recovery of Rhetoric: Persuasive Discourse and Disciplinarity in the Human Sciences* (London: Bristol Classical Press, Duckworth, 1993).

—— 'Disciplining social psychology: a case study of boundary relations in the history of the human sciences', *Journal of the History of the Behavioral Sciences*, 36 (2000), 383–403.

Gould, Stephen Jay, *The Mismeasurement of Man* (1981) (Harmondsworth: Penguin, 1984).

Gouldner, Alvin W., *The Coming Crisis of Western Sociology* (1970) (London: Heinemann, 1971).

Greenfield, Susan, *Journey to the Centers of the Mind: Toward a Science of Consciousness* (New York: W. H. Freeman, 1995).

—— 'Soul, brain and mind', in M. James C. Crabbe (ed.), *From Soul to Self* (London: Routledge, 1999).

Groethuysen, Bernard, 'Towards an anthropological philosophy', trans. Sheila A. Kerr, in Raymond Klibansky and H. J. Paton (eds), *Philosophy & History: Essays Presented to Ernst Cassirer* (Oxford: Oxford University Press, 1936).

Gusdorf, Georges, *Introduction aux sciences humaines: essai critique sur leurs origines et leur développement*, Publications de la Faculté des Lettres de l'Université de Strasbourg, fascicule 140 (Paris: En Dépôt à la Société d'Edition, Les belles lettres, 1960).

—— *De l'histoire des sciences a l'histoire de la pensée: les sciences humaines et la pensée occidentale I* (1966) (Paris: Payot, 1977).

—— *Les Sciences humaines et la pensée occidentale*, 12 vols (Paris: Payot, 1966–85).

Habermas, Jürgen, *Knowledge and Human Interests* (1968), trans. Jeremy J. Shapiro (London: Heinemann, 1972).

Hacking, Ian, *Historical Ontology* (Cambridge, Mass.: Harvard University Press, 2002).

Hallberg, Margareta, Bengt Molander and Lennart Olausson, *Reflections of Humanities Studies* (Göteborg: Humanities Studies Series, Faculty of Humanities, Göteborgs Universitet, 1998).

Hampshire, Stuart, *Thought and Action* (1959) (New York: Viking Press, 1960).

Haraway, Donna, *Primate Visions: Gender, Race, and Nature in the World of Modern Science* (New York: Routledge, 1990).

Harré, R. and P. F. Secord, *The Explanation of Social Behaviour* (Oxford: Basil Blackwell, 1972).

Hayward, Rhodri, '"Our friends electric": mechanical models of mind in postwar Britain', in G. C. Bunn, A. D. Lovie and G. D. Richards (eds), *Psychology in Britain: Historical Essays and Personal Reflections* (Leicester: British Psychological Society, 2001).

Hegel, G. W. F., *Phenomenology of Spirit* (1807), trans. A. V. Miller (Oxford: Oxford University Press, 1977).

Heidegger, Martin, *Being and Time* (1927), trans. John Macquarrie and Edward Robinson (Oxford: Basil Blackwell, 1967).

Heilbron, Johan, *The Rise of Social Theory* (1990), trans. Sheila Gogol (Cambridge: Polity Press, 1995).

Helmholtz, Hermann von, 'The relation of the natural sciences to science in general' (lecture 1862), trans. H. W. Eve, revised Russell Kahl, in Russell Kahl (ed.), *Selected Writings of Hermann von Helmholtz* (Middletown, Wesleyan University Press, 1971).

Hennis, Wilhelm, 'Max Weber's "central question"', trans. Keith Tribe, *Economy and Society*, 12 (1983), 135–80.

Herder, Johann Gottfried von, 'On cognition and sensation, the two main forces of the human soul' (preface from draft written 1775), in Michael N. Forster (ed. and trans.), *Philosophical Writings* (Cambridge: Cambridge University Press, 2002).

——— 'On cognition and sensation of the human soul' (1778), in Michael N. Forster (ed. and trans.), *Philosophical Writings* (Cambridge: Cambridge University Press, 2002).

——— *Reflections on the Philosophy of the History of Mankind* (1784–91), trans. T. O. Churchill (1800), abridged Frank E. Manuel (Chicago: University of Chicago Press, 1968).

Hesse, Mary, 'Theory and observation' (1970), in *Revolutions and Reconstructions in the Philosophy of Science* (Brighton: Harvester Press, 1980).

Hoffmann, E. T. A., 'The choosing of the bride' (revised 1820), trans. R. J. Hollingdale, in *Tales of Hoffmann* (London: Penguin Books, 2004).

Honneth, Axel and Hans Joas, *Social Action and Human Nature* (1980), trans. Raymond Meyer (Cambridge: Cambridge University Press, 1988).

Humboldt, Wilhelm von, 'On the historian's task' (lecture 1821), trans. anon., *History and Theory*, 6 (1967), 57–71.

——— *On Language: On the Diversity of Human Language Construction and Its Influence on the Mental Development of the Human Species* (1836), ed. Michael Losonsky, trans. Peter Heath (Cambridge: Cambridge University Press, 1999).

Hume, David, *A Treatise of Human Nature* (1739–40), ed. L. S. Selby-Bigge (Oxford: Clarendon Press, 1888).

——— 'Of natural characters' (1748), in *Essays and Treatises on Several Subjects*, vol. 1, *Essays, Moral, Political, and Literary* (Bristol: Thoemmes Press, 2002).

——— *Enquiries Concerning the Human Understanding and Concerning the Principles of Morals* (new edn 1777), ed. L. A. Selby-Bigge, 2nd edn (Oxford: Clarendon Press, 1902).

Husserl, Edmund, 'The crisis of the sciences as expression of the radical life-crisis of European humanity' (1936), in *The Crisis of European Sciences and Transcendental Phenomenology: An Introduction to Phenomenological Philosophy*, trans. David Carr (Evanston: Northwestern University Press, 1970).

Izenberg, Gerald N., *The Existentialist Critique of Freud: The Crisis of Autonomy* (Princeton: Princeton University Press, 1976).

Kant, Immanuel, 'An answer to the question: "What is enlightenment?"' (1784), in Hans Reiss (ed.), H. B. Nisbet (trans.), *Kant's Political Writings* (Cambridge: Cambridge University Press, 1970).

—— 'Idea for a universal history with a cosmopolitan purpose' (1784), in Hans Reiss (ed.), H. B. Nisbet (trans.), *Kant's Political Writings* (Cambridge: Cambridge University Press, 1970).

—— *Critique of Judgment. Including the First Introduction* (1790), trans. and intro. Werner S. Pluhar (Indianapolis: Hackett, 1987).

—— *Anthropology from a Pragmatic Point of View* (2nd German edn 1800), trans. Mary J. Gregor (The Hague: Martinus Nijhoff, 1974).

—— *Immanuel Kant's Logic: A Manual for Students*, ed. Gottlob Benjamin Jäsche (1800), in J. Michael Young (ed.), *Lectures on Logic* (Cambridge: Cambridge University Press, 1992).

—— *Lectures on Metaphysics*, ed. and trans. Karl Ameriks and Steve Naragon (Cambridge: Cambridge University Press, 1997).

Kay, Lily E., 'Rethinking institutions: Philanthropy as an historiographic problem of knowledge and power', *Minerva*, 35 (1997), 283–93.

Kelley, Donald R., *The Human Measure: Social Thought in the Western Legal Tradition* (Cambridge, Mass.: Harvard University Press, 1990).

—— 'The problem of knowledge and the concept of discipline', in Donald R. Kelley (ed.), *History and the Disciplines: The Reclassification of Knowledge in Early Modern Europe* (Rochester, N.Y.: University of Rochester Press, 1997).

Kelly, Aileen M., *Toward Another Shore: Russian Thinkers between Necessity and Chance* (Hew Haven: Yale University Press, 1998).

Kolakowski, Leszek, *Positivist Philosophy: From Hume to the Vienna Circle* (1966), trans. Norbert Guterman (Harmondsworth: Penguin Books, 1972).

Krohn, Wolfgang, Günther Küppers and Helga Nowotny (eds), *Selforganization: Portrait of a Scientific Revolution*, Sociology of the Sciences Yearbook, vol. 14 (Dordrecht: Kluwer, 1990).

Kuhn, Thomas S., 'The function of measurement in modern physical science' (1961), in *The Essential Tension: Selected Studies in Scientific Tradition and Change* (Chicago: University of Chicago Press, 1977).

—— 'The relations between history and history of science' (1971), in *The Essential Tension: Selected Studies in Scientific Tradition and Change* (Chicago: University of Chicago Press, 1977).

Kuper, Adam, *The Chosen Primate: Human Nature and Cultural Diversity* (Cambridge, Mass.: Harvard University Press, 1994).

—— 'On human nature: Darwin and the anthropologists', in Mikuláš Teich, Roy Porter and Bo Gustafsson (eds), *Nature and Society in Historical Context* (Cambridge: Cambridge University Press, 1997).

Kusch, Martin, *Psychological Knowledge: A Social History and Philosophy* (London: Routledge, 1999).

Langer, Susanne K., *Philosophy in a New Key: A Study in the Symbolism of Reason, Rite, and Art* (1942), 3rd edn (Cambridge, Mass.: Harvard University Press, 1957).

Latour, Bruno, *The Pasteurization of France* (1984), trans. Alan Sheridan and John Law (Cambridge, Mass.: Harvard University Press, 1988).

—— *Pandora's Hope: Essays on the Reality of Science Studies* (Cambridge, Mass.: Harvard University Press, 1999).

Lawson, Hilary, *Reflexivity: The Post-Modern Predicament* (London: Hutchinson, 1985).

Leterrier, Sophie-Anne, *L'Institution des sciences morales: l'Académie des Sciences Morales et Politiques 1795–1850* (Paris: L'Harmattan, 1995).

Lewes, George Henry, 'Spiritualism and materialism', *Fortnightly Review*, new series 19 (1876), 479–93, 707–19.

Lloyd, Genevieve, 'Maleness, metaphor, and the "crisis" of reason', in Louise M. Anthony and Charlotte E. Witt (eds), *A Mind of One's Own: Feminist Essays on Reason and Objectivity* (1992), 2nd edn (Boulder: Westview Press, 2002).

Locke, John, *An Essay Concerning Human Understanding* (1690), ed. John W. Yolton, revised edn, 2 vols (London: Dent, 1965).

McDonald, Henry, 'Language and being: crossroads of modern literary theory and classical ontology', *Philosophy & Social Criticism*, 30:2 (2004), 187–220.

MacIntyre, Alasdair, 'A mistake about causality in social science', in Peter Laslett and W. G. Runciman (eds), *Philosophy, Politics and Society (Second Series)* (Oxford: Basil Blackwell, 1962).

—— 'Epistemological crises, dramatic narrative and the philosophy of science', *The Monist*, 60 (1977), 453–72.

—— *After Virtue: A Study in Moral Theory* (London: Duckworth, 1981).

Malik, Kenan, *The Meaning of Race: Race, History and Culture in Western Society* (Basingstoke: Macmillan, 1996).

—— *Man, Beast and Zombie: What Science Can and Cannot Tell Us about Human Nature* (London: Weidenfeld & Nicolson, 2000).

Mandelbaum, Maurice, *History, Man, & Reason: A Study in Nineteenth-Century Thought* (Baltimore: The Johns Hopkins University Press, 1971).

Manicas, Peter T., *A History and Philosophy of the Social Sciences* (Oxford: Basil Blackwell, 1987).

Margolis, Joseph, *Science without Unity: Reconciling the Human and Natural Sciences* (Oxford: Basil Blackwell, 1987).

—— *The Flux of History and the Flux of Science* (Berkeley: University of California Press, 1993).

Markus, Gyorgy, 'Why is there no hermeneutics of natural sciences? Some preliminary theses', *Science in Context*, 1 (1987), 5–51.

Marx, Karl, 'Economic and philosophical manuscripts' (written 1844), trans. Gregor Benton, in *Early Writings*, intro. Lucio Colletti (London: Penguin Books, 1995).

—— *Capital: A Critique of Political Economy*, vol. 1 (3rd German edn 1883), trans. Samuel Moore and Edward Aveling, ed. Frederick Engels (London: Lawrence & Wishart, 1954).

Marx, Karl and Frederick Engels, *The German Ideology. Parts I and III* (written 1845–46), ed. R. Pascal, trans. anon. (New York: International Publishers, 1947).

Maturana, Humberto R., 'Biology of cognition' (1970), in Humberto R. Maturana and Francisco J. Varela, *Autopoiesis and Cognition: The Realization of the Living* (Dordrecht: D. Reidel, 1980).

Maturana, Humberto R. and Francisco J. Varela, 'Autopoiesis: The organization of the living' (1972), in *Autopoiesis and Cognition: The Realization of the Living* (Dordrecht: D. Reidel, 1980).

—— *The Tree of Knowledge: The Biological Roots of Human Understanding* (1987), revised edn (Boston: Shambhala, 1992).

Mayr, Ernst, *The Growth of Biological Thought: Diversity, Evolution, and Inheritance* (Cambridge, Mass.: Harvard University Press, 1982).

Mazlish, Bruce, *The Uncertain Sciences* (New Haven: Yale University Press, 1998).

Mead, George Herbert, 'Social psychology as counterpart to physiological psychology' (1909), in Andrew J. Reck (ed.), *Selected Writings* (Indianapolis: Bobbs-Merrill, 1964).

—— 'What social objects must psychology presuppose?' (1910), in Andrew J. Reck (ed.), *Selected Writings* (Indianapolis: Bobbs-Merrill, 1964).

Medawar, P. B., *The Art of the Soluble* (London: Methuen, 1967).

Megill, Allan, *Prophets of Extremity: Nietzsche, Heidegger, Foucault, Derrida* (Berkeley: University of California Press, 1985).

—— (ed.), *Rethinking Objectivity* (Durham: Duke University Press, 1994).

Midgley, Mary, *Beast and Man: The Roots of Human Nature* (Ithaca: Cornell University Press, 1978).

—— *Science as Salvation: A Modern Myth and Its Meaning* (London: Routledge, 1992).

Mill, John Stuart, *A System of Logic Ratiocinative and Inductive: Being a Connected View of the Principles of Evidence and the Methods of Scientific Investigation* (1843), in J. M. Robson (ed.), *Collected Works of John Stuart Mill*, vol. 8 (Toronto: University of Toronto Press, and London: Routledge & Kegan Paul, 1974).

Mink, Louis O., 'Narrative form as a cognitive instrument', in Robert H. Canary and Henry Kozicki (eds), *The Writing of History: Literary Form and Historical Understanding* (Madison: University of Wisconsin Press, 1978).

Monod, Jacques, *Chance and Necessity: An Essay on the Natural Philosophy of Modern Biology* (1970), trans. Austryn Wainhouse (London: Collins, 1972).

Morawski, Jill G., 'The measurement of masculinity and femininity: engendering categorical realities', *Journal of Personality*, 53 (1985), 196–223.

—— 'Self-regard and other-regard: Reflexive practices in American psychology, 1890–1940', *Science in Context*, 5 (1992), 281–308.

—— *Practicing Feminisms, Reconstructing Psychology: Notes on a Liminal Science* (Ann Arbor: University of Michigan Press, 1994).

—— 'Reflexivity and the psychologist', *History of the Human Sciences*, 18:4 (2005), 77–105.

[Mozley, J. R.], 'Philosophy, psychology, and metaphysics', *North British Review*, 14 (1870), 115–39.

Müller-Sievers, Helmut, *Self-Generation: Biology, Philosophy, and Literature around 1800* (Stanford: Stanford University Press, 1997).

Murdoch, Iris, 'Metaphysics and ethics' (1957), in *Existentialists and Mystics: Writings on Philosophy and Literature* (New York: Allen Lane, Penguin Press, 1998).

Myers, Greg, *Writing Biology: Texts in the Social Construction of Scientific Knowledge* (Madison: University of Wisconsin Press, 1990).

Nagel, Thomas, *The View from Nowhere* (New York: Oxford University Press, 1986).

Needham, Joseph, 'A biologist's view of Whitehead's philosophy' (1941), in *Time: The Refreshing River (Essays and Addresses, 1932–1942)* (London: George Allen & Unwin, 1943).

Nelson, John S., Allan Megill and Donald N. McCloskey, 'Rhetoric of inquiry', in John S. Nelson, Allan Megill and Donald N. McCloskey (eds), *The Rhetoric of the Human Sciences: Language and Argument in Scholarship and Public Affairs* (Madison: University of Wisconsin Press, 1987).

—— (eds), *The Rhetoric of the Human Sciences: Language and Argument in Scholarship and Public Affairs* (Madison: University of Wisconsin Press, 1987).

Nietzsche, Friedrich, *On the Genealogy of Morals* (1887), ed. Walter Kaufmann, trans. Walter Kaufmann and R. J. Hollingdale (New York: Vintage Books, 1969).

Oakes, Guy, *Weber and Rickert: Concept Formation in the Cultural Sciences* (Cambridge, Mass.: MIT Press, 1988).

O'Hear, Anthony, '"Two cultures" revisited', in Anthony O'Hear (ed.), *Verstehen and Humane Understanding*, Royal Institute of Philosophy Supplement, 41 (Cambridge: Cambridge University Press, 1996).

Ortega y Gasset, José, 'History as a system', trans. William C. Atkinson, in Raymond Klibansky and H. J. Paton (eds), *Philosophy & History: Essays Presented to Ernst Cassirer* (Oxford: Clarendon Press, 1936).

Pagden, Anthony, 'Eighteenth-century anthropology and the "history of mankind"', in Donald R. Kelley (ed.), *History and the Disciplines: The Reclassification of Knowledge in Early Modern Europe* (Rochester, N.Y.: University of Rochester Press, 1997).

Pascal, Blaise, *Pensées* (1670), trans. and intro. A. J. Krailsheimer, revised edn (London: Penguin Books, 1995).

Pickering, Andy, 'Cyborg history and the World War II regime', *Perspectives on Science*, 3 (1995), 1–48.

Pinkard, Terry, *German Philosophy 1760–1860: The Legacy of Idealism* (Cambridge: Cambridge University Press, 2002).

Pinker, Steven, *How the Mind Works* (1997) (London: Penguin Books, 1998).

Pleasants, Nigel, 'The concept of learning from the study of the Holocaust', *History of the Human Sciences*, 17:2–3 (2004), 187–210.

Polanyi, Michael, *Personal Knowledge: Towards a Post-Critical Philosophy* (London: Routledge & Kegan Paul, 1958).

Porter, Theodore M., and Dorothy Ross, 'Introduction: writing the history of social science', in Theodore M. Porter and Dorothy Ross (eds), *The Cambridge History of Science, vol. 7, The Modern Social Sciences* (Cambridge: Cambridge University Press, 2003).

Putnam, Hilary, 'The place of facts in a world of values' (1976), in James Conant (ed.), *Realism with a Human Face* (Cambridge, Mass.: Harvard University Press, 1990).

—— *Reason, Truth and History* (Cambridge: Cambridge University Press, 1981).

—— 'Beyond historicism' (1981), in *Realism and Reason: Philosophical Papers*, vol. 3 (Cambridge: Cambridge University Press, 1983).

—— 'Beyond the fact/value dichotomy' (1982), in James Conant (ed.), *Realism with a Human Face* (Cambridge, Mass.: Harvard University Press, 1990).

—— *The Collapse of the Fact/Value Dichotomy and Other Essays* (Cambridge, Mass.: Harvard University Press, 2002).

Queneau, Raymond, *Exercises in Style* (1947), trans. Barbara Wright (London: John Calder, 1979).

Randall, Jr, John Herman, 'Epilogue: the nature of naturalism', in Yervant H. Krikorian (ed.), *Naturalism and the Human Spirit* (New York: Columbia University Press, 1944).

Richards, Graham, 'Of what is the history of psychology a history?', *British Journal for the History of Science*, 20 (1987), 201–11.

—— *Putting Psychology in Its Place: A Critical Historical Overview* (1996), 2nd edn (London: Routledge/Psychology Press, 2002).

—— *'Race', Racism and Psychology: Towards a Reflexive History* (London: Routledge, 1997).

Richards, Robert J., *The Romantic Conception of Life: Science and Philosophy in the Age of Goethe* (Chicago: University of Chicago Press, 2002).

Rickert, Heinrich, *Science and History: A Critique of Positivist Epistemology* (6th and 7th German edns 1926), ed. Arthur Goddard, trans. George Reisman (Princeton: D. Van Nostrand, 1962).

—— *The Limits of Concept Formation in Natural Science: A Logical Introduction to the Historical Sciences* (5th German edn 1929), abridged, ed. and trans. Guy Oakes (Cambridge: Cambridge University Press, 1986).

Ricoeur, Paul, 'What is a text? Explanation and understanding' (1970), in John B. Thompson (ed. and trans.), *Hermeneutics and the Human Sciences: Essays on Language, Action and Interpretation* (Cambridge: Cambridge University Press, 1981).

—— 'The narrative function' (1979), in John B. Thompson (ed. and trans.), *Hermeneutics and the Human Sciences: Essays on Language, Action and Interpretation* (Cambridge: Cambridge University Press, 1981).

—— *Time and Narrative*, vol. 1 (1983), trans. Kathleen McLaughlin and David Pellauer (Chicago: University of Chicago Press, 1984).

Ringer, Fritz K., *The Decline of the German Mandarins: The German Academic Community, 1890–1933* (1969) (Hanover, N.H.: Wesleyan University Press/University Presses of New England, 1990).

Ritter, Joachim (ed.), 'Geisteswissenschaften', in *Historisches Wörterbuch der Philosophie*, vol. 3 (Darmstadt: Wissenschaftliche Buchgesellschaft, 1974).

Roberts, Richard H., and James M. M. Good (eds), *The Recovery of Rhetoric: Persuasive Discourse and Disciplinarity in the Human Sciences* (London: Bristol Classical Press, Duckworth, 1993).

Rorty, Richard, *Philosophy and the Mirror of Nature* (1979) (Oxford: Basil Blackwell, 1980).

—— *Contingency, Irony, and Solidarity* (Cambridge: Cambridge University Press, 1989).

Rose, Steven (ed.), *From Brains to Consciousness? Essays on the New Sciences of the Mind* (London: Allen Lane, Penguin Press, 1998).

Ross, Dorothy, 'Changing contours of the social science disciplines', in Theodore M. Porter and Dorothy Ross (eds), *The Cambridge History of Science, vol. 7, The Modern Social Sciences* (Cambridge: Cambridge University Press, 2003).

Rouse, Joseph, 'The narrative reconstruction of science', *Inquiry*, 33 (1990), 179–96.

Ruse, Michael, *Monad to Man: The Concept of Progress in Evolutionary Biology* (Cambridge, Mass.: Harvard University Press, 1996).

Samelson, Franz, 'From "race psychology" to "studies in prejudice": some observations on the thematic reversal in social psychology', *Journal of the History of the Behavioral Sciences*, 14 (1978), 265–78.

Sandywell, Barry, *Reflexivity and the Crisis of Western Reason: Logological Investigations Volume 1* (London: Routledge, 1996).

Sartre, Jean-Paul, *Being and Nothingness: An Essay on Phenomenological Ontology* (1943), trans. Hazel E. Barnes (London: Methuen, 1958).

Sass, Louis A., *Madness and Modernism: Insanity in the Light of Modern Art, Literature, and Thought* (Cambridge, Mass.: Harvard University Press, 1992).

Schnädelbach, Herbert, *Philosophy in Germany 1831–1933* (1983), trans. Eric Matthews (Cambridge: Cambridge University Press, 1984).

Schneewind, J. B., 'No discipline, no history: the case of moral philosophy', in Donald R. Kelley (ed.), *History and the Disciplines: The Reclassification of Knowledge in Early Modern Europe* (Rochester, N.Y.: University of Rochester Press, 1997).

Searle, John R., *The Rediscovery of the Mind* (Cambridge, Mass.: MIT Press, 1992).

Seigel, Jerrold, *The Idea of the Self: Thought and Experience in Western Europe since the Seventeenth Century* (Cambridge: Cambridge University Press, 2005).

Shaffer, Elinor R., 'Romantic philosophy and the organization of the disciplines: the founding of the Humboldt University of Berlin', in

Andrew Cunningham and Nicholas Jardine (eds), *Romanticism and the Sciences* (Cambridge: Cambridge University Press, 1990).

Shamdasani, Sonu, *Jung and the Making of Modern Psychology: The Dream of a Science* (Cambridge: Cambridge University Press, 2003).

Shapin, Steven, *The Scientific Revolution* (Chicago: University of Chicago Press, 1996).

—— 'How to be antiscientific', in Jay A. Labinger and Harry Collins (eds), *The One Culture? A Conversation about Science* (Chicago: University of Chicago Press, 2001).

Shelley, Percy Bysshe, 'A defence of poetry' (written 1821), in Donald H. Reiman and Neil Fraistat (eds), *Shelley's Poetry and Prose: Authoritative Texts: Criticism*, 2nd edn (New York: W. W. Norton, 2002).

Sherrington, Charles, *Man on His Nature* (Cambridge: Cambridge University Press, 1940).

Sirotkina, Irina, 'A family discussion: the Herzens on the science of man', *History of the Human Sciences*, 15:4 (2002), 3–21.

Skinner, Quentin, 'Meaning and understanding in the history of ideas' (1969), in James Tully (ed.), *Meaning and Context: Quentin Skinner and His Critics* (Cambridge: Polity Press, 1988).

Smith, Adam, *An Inquiry into the Nature and Causes of the Wealth of Nations* (1776), ed. R. H. Campbell and A. S. Skinner, 2 vols (Oxford: Clarendon Press, 1976).

Smith, Barbara Herrnstein, *Belief and Resistance: Dynamics of Contemporary Intellectual Controversy* (Cambridge, Mass.: Harvard University Press, 1997).

Smith, Roger, 'Mad or bad? Victorian stories of the criminally insane', *LSE Quarterly*, 3 (1989), 1–20.

—— *Inhibition: History and Meaning in the Sciences of Mind and Brain* (London: Free Association Books, and Berkeley: University of California Press, 1992).

—— *The Fontana History of the Human Sciences* (London: Fontana, 1997), same as *The Norton History of the Human Sciences* (New York: W. W. Norton, 1997).

—— 'The big picture: writing psychology into the history of the human sciences', *Journal of the History of the Behavioral Sciences*, 34 (1998), 1–14.

—— '"Nauka o cheloveke" i moral'naya filosofiya: otnoshenie faktov i tsennostei', trans. Irina Sirotkina, in M. T. Stepanyants (ed.), *Sravnitel'naya filosofiya: Moral'naya filosofiya v kontekste mnogoobraziya kul'tur* (Moscow: 'Vostochnaya Literatura', Russian Academy of Sciences, 2004).

—— 'Does reflexivity separate the human sciences from the natural sciences?', *History of the Human Sciences*, 18:4 (2005), 1–25.

—— 'The history of human nature: more of the same or facing the other?' in Asya Syrodeeva (ed.), *Kollazh – 5: Sotsial'no-filosofskii i filosofsko-antropologicheskii al'manakh* (Moscow: Institute of Philosophy, Russian Academy of Sciences, 2005).

—— 'The history of psychological categories', *Studies in History and Philosophy of Biological and Biomedical Sciences*, 36 (2005), 55–94.

—— (ed.), 'Reflexivity', special issue, *History of the Human Sciences*, 18:4 (2005).

Staeuble, Irmingard, '"Psychological man" and human subjectivity in historical perspective', *History of the Human Sciences*, 4 (1991), 417–32.

Starobinski, Jean, 'The critical relation' (1970), in *The Living Eye*, trans. Arthur Goldhammer (Cambridge, Mass.: Harvard University Press, 1989).

—— *Action and Reaction: The Life and Adventures of a Couple* (1999), trans. Sophie Hawkes (New York: Zone Books, 2003).

Steedman, Carolyn, *Dust* (Manchester: Manchester University Press, 2001).

Stichweh, Rudolf, 'The sociology of scientific disciplines: on the genesis and stability of the disciplinary structure of modern science', *Science in Context*, 5 (1992), 3–15.

Still, A. W., and J. M. M. Good, 'The ontology of mutualism', *Ecological Psychology*, 10:1 (1998), 39–63.

Stone, Lawrence, 'The revival of narrative: reflections on a new old history', *Past and Present*, 85 (1979), 1–24.

Tauber, Alfred I., *Confessions of a Medicine Man: An Essay in Popular Philosophy* (Cambridge, Mass.: MIT Press, 1999).

—— *Henry David Thoreau and the Moral Agency of Knowing* (Berkeley: University of California Press, 2001).

Tauber, Alfred I. (ed.), *Science and the Quest for Reality* (New York: New York University Press, 1997).

Taylor, Charles, 'How is mechanism conceivable?' (1971), in *Human Agency and Language: Philosophical Papers 1* (Cambridge: Cambridge University Press, 1985).

—— 'Interpretation and the sciences of man' (1971), in *Philosophy and the Human Sciences: Philosophical Papers 2* (Cambridge: Cambridge University Press, 1985).

—— *Hegel* (Cambridge: Cambridge University Press, 1975).

—— *Human Agency and Language: Philosophical Papers 1* (Cambridge: Cambridge University Press, 1985).

—— *Philosophy and the Human Sciences: Philosophical Papers 2* (Cambridge: Cambridge University Press, 1985).

—— *Sources of the Self: The Making of the Modern Identity* (Cambridge: Cambridge University Press, 1989).

—— 'The importance of Herder' (1991) in *Philosophical Arguments* (Cambridge, Mass.: Harvard University Press, 1995).

Toews, John Edward, *Hegelianism: The Path towards Dialectical Humanism, 1805–1841* (Cambridge: Cambridge University Press, 1980).

Tooby, John and Leda Cosmides, 'The psychological foundations of culture', in Jerome H. Barkow, Leda Cosmides and John Tooby (eds), *The Adapted Mind: Evolutionary Psychology and the Generation of Culture* (New York: Oxford University Press, 1992).

Trigg, Roger, *The Shaping of Man: Philosophical Aspects of Sociobiology* (Oxford: Basil Blackwell, 1982).

—— *Ideas of Human Nature: An Historical Introduction* (1988), 2nd edn (Oxford: Blackwell, 1999).

Turkle, Sherry, *The Second Self: Computers and the Human Spirit* (New York: Simon and Schuster, 1984).

Vico, Giambattista, *New Science: Principles of the New Science Concerning the Common Nature of Nations* (3rd edn 1744), trans. David Marsh (London: Penguin Books, 1999).

Wagner, Peter, *A History and Theory of the Social Sciences: Not All that Is Solid Melts into Air* (London: Sage, 2001).

Wagner, Peter, Björn Wittrock and Richard Whitley (eds), *Discourses on Society: The Shaping of the Social Science Disciplines*, Sociology of the Sciences Yearbook, vol. 15 (Dordrecht: Kluwer, 1991).

Wallace, Alfred Russel, 'The origin of human races and the antiquity of man deduced from the theory of "natural selection"' (1864), in Michael D. Biddiss (ed.), *Images of Race* (New York: Holmes & Meier, 1979).

Wartofsky, Marx W., 'Perception, representation, and the forms of action: towards an historical epistemology' (1973), in Robert S. Cohen and Marx W. Wartofsky (eds), *A Portrait of Twenty-Five Years: Boston Colloquium for the Philosophy of Science 1960–1985* (Dordrecht: D. Riedel, 1985).

Weber, Max, '"Objectivity" in social science and social policy' (1904), in Edward A. Shils and Henry A. Finch (ed. and trans.), *The Methodology of the Social Sciences* (New York: Free Press, 1949).

—— 'The meaning of "ethical neutrality" in sociology and economics' (1917), in Edward A. Shils and Henry A. Finch (ed. and trans.), *The Methodology of the Social Sciences* (New York: Free Press, 1949).

—— 'Science as a vocation' (1919), trans. Michael John, in Peter Lassman and Irving Velody (eds), with Herminio Martins, *Max Weber's 'Science as a Vocation'* (London: Unwin Hyman, 1989).

—— *Economy and Society: An Outline of Interpretive Sociology* (4th German edn 1956), ed. Guenther Roth and Claus Wittock, trans. Ephraim Fischoff et al., 2 vols (Berkeley: University of California Press, 1978).

Whewell, William, *The Philosophy of the Inductive Sciences, Founded upon Their History* (1840), 2nd edn, 2 vols (London: John W. Parker, 1847).

White, Hayden, *Metahistory: The Historical Imagination in Nineteenth-Century Europe* (Baltimore: The Johns Hopkins University Press, 1973).

—— *The Content of the Form: Narrative Discourse and Historical Representation* (Baltimore: The Johns Hopkins University Press, 1987).

—— 'Commentary', special issue, intro. Irving Velody, 'Identity, memory and history', *History of the Human Sciences*, 9:4 (1996), 123–38.

Whitehead, A. N., *Science and the Modern World* (1926) (Cambridge: Cambridge University Press, 1953).

—— *Adventures of Ideas* (New York: Macmillan, 1933).

—— *Modes of Thought* (Cambridge: Cambridge University Press, 1938).

Wilson, Edward O., *Sociobiology: The New Synthesis* (Cambridge, Mass.: Harvard University Press, 1975).

Winch, Peter, *The Idea of a Social Science and Its Relation to Philosophy* (1958), corrected edn (London: Routledge & Kegan Paul, 1963).

Windelband, Wilhelm, 'History and natural science' (rectorial address 1894), trans. Guy Oakes, *History and Theory*, 19 (1980), 165–86.

Wittgenstein, Ludwig, *Tractatus Logico-Philosophicus* (1921), trans. D. F. Pears and B. F. McGuinness (London: Routledge & Kegan Paul, 1961).

—— *Philosophical Investigations* (1953), trans. G. E. M. Anscombe, 2nd edn (Oxford: Basil Blackwell, 1958).

—— *On Certainty*, ed. G. E. M. Anscombe and G. H. von Wright, trans. Denis Paul and G. E. M. Anscombe (Oxford: Basil Blackwell, 1969).

Wittrock, Björn, Johan Heilbron and Lars Magnusson, 'The rise of the social sciences and the formation of modernity', in Johan Heilbron, Lars Magnusson and Björn Wittrock (eds), *The Rise of the Social Sciences and the Formation of Modernity: Conceptual*

*Change in Context, 1750–1850*, Sociology of the Sciences Yearbook, vol. 20 (Dordrecht: Kluwer, 1998).

Wokler, Robert, 'Anthropology and conjectural history in the Enlightenment', in Christopher Fox, Roy Porter and Robert Wokler (eds), *Inventing Human Science: Eighteenth-Century Domains* (Berkeley: University of California Press, 1995).

—— 'The Enlightenment and the French revolutionary birth pangs of modernity', in Johan Heilbron, Lars Magnusson and Björn Wittrock (eds), *The Rise of the Social Sciences and the Formation of Modernity: Conceptual Change in Context, 1750–1850*, Sociology of the Sciences Yearbook, vol. 20 (Dordrecht: Kluwer, 1998).

Yeo, Richard, *Defining Science: William Whewell, Natural Knowledge, and Public Debate in Early Victorian Britain* (Cambridge: Cambridge University Press, 1993).

Young, Robert M., *Mind, Brain, and Adaptation in the Nineteenth Century: Cerebral Localization and Its Biological Context from Gall to Ferrier* (Oxford: Clarendon Press, 1970).

—— *Darwin's Metaphor: Nature's Place in Victorian Culture* (Cambridge: Cambridge University Press, 1985).

Zammito, John H., *Kant, Herder, and the Birth of Anthropology* (Chicago: University of Chicago Press, 2002).

# Index

'n.' after a page reference indicates a note on that page.

Abrams, M. H. 139n.29, 258n.42
Académie des Sciences Morales et
    Politiques 141–2
Adorno, Theodor 241
agency, human 4–5, 98–107,
    189–90, 236
Anderson, Perry 24–5
Anthony, Louise M. 84n.61
'anthropological turn' 53, 54
anthropomorphism 42, 197
Aquinas, St Thomas 26
Arbib, Michael A. 113, 226–7, 235,
    239
Aristotle 139
    *De Anima* 140
Azouvi, François 140n.29

Bacon, Francis 131
Baker, Keith Michael 87, 141n.30,
    142n.34
Barendregt, Johan T. 215n.37
Barham, Peter 188n.37
Barkow, Jerome H. 219n.43
Bazerman, Charles 180n.21
Beckner, Morton 226n.58
Beer, Gillian 177, 202
behavioral sciences 98, 212
Ben-Chaim, Michael 229n.63
Berlin, Isaiah 10, 75, 105, 109,
    120, 133n.19, 134, 136n.22,
    246
Bernal, J. D. 71

Bernstein, Richard J. 75n.34, 95,
    198, 208, 210
Bijker, Wiebe E. 70n.20
*Bildung* 39, 51, 168–9
Billig, Michael 115
Blanckaert, Claude 252n.26
Blondiaux, Loïc 183n.27, 216n.38
Blumenberg, Hans 133
Bouwsma, William J. 176n.9
Boyle, Nicholas 43n.45
brain 3, 33–4, 108, 226
    scanning 113–14
Braunstein, Jean-François 21n.6,
    94n.1, 107n.26
Breuer, Josef, and talking cure 185
Brixey, Martha, story of 185–91
Buckle, H. T. 127, 149
Buffon, comte de, *Histoire naturelle*
    141
Burnham, John C. 124n.4
Burrow, John 21n.9, 148n.42
Burtt, E. A. 236–8, 240

Calame, Claude 252
Callon, Michel 70
Canguilhem, Georges 20, 107, 142
Carr, David 175
Carrithers, David 86
Carroy, Jacqueline 85n.64
Cartesian 'ego' 57, 63, 79
Cartesian separation of mind and
    body 140, 236–9

Cassirer, Ernst 49–51, 54, 115, 161, 225–6, 245–6, 255–7
  *Essay on Man* 63
  form-analysis 164–6
Certeau, Michel de 179, 230, 234, 239
Churchland, Paul M. 33, 108n.27, 109n.29
Clark, William 36n.35
Clifford, W. K. 95
Coleman, Janet 195
Collingwood, R. G. 9, 102–3, 115–16, 119, 120, 176, 178–9, 181, 192, 232, 244, 258
Collini, Stefan 21n.9, 25, 148n.42
Comte, Auguste 2–3, 64, 86–7, 121, 142
  *Cours de philosophie positive* 93–4
Condorcet, marquis de 142
consciousness, material explanation of 33–4, 108–9, 114, 256
Cosmides, Lena 28–9
  'Psychological foundations of culture' 9n.12, 218–22
Coulter, Jeff 111n.30
Crick, Francis 11
crime and insanity 186–9, 227
cultural crisis 47, 148–50, 157–8, 163, 167, 207
cultural sciences 105–6, 125–6
culture, definition of 34–5, 40, 50, 72–5, 103, 106, 160

Danto, Arthur C. 176–7
Danziger, Karl 77n.39, 81n.54
Darwin, Charles 9, 28, 192, 218, 236–7
  *The Descent of Man* 30
  evolution of morality 30–2
Daston, Lorraine 224n.53
Dawkins, Richard 7, 11, 249
  *The Selfish Gene* 200–5
Dehue, Trudy 215n.37
Dennett, Daniel 7
  *Consciousness Explained* 34
Derrida, Jacques 7, 59, 66, 69
Descartes, René 63, 79, 84, 112, 131, 135, 140, 238

desire 36–7, 185, 254–6
Destutt de Tracy, comte A.-L.-C. 142
Dewey, John 51, 92, 221, 227, 246
Diderot, Denis, *Encyclopédie* 20
Dilthey, Wilhelm 14, 45, 52, 55, 60, 127, 128, 130, 149, 150, 151–8, 160, 162–4, 166, 169, 184, 195, 228
  *Introduction to the Human Sciences* 123–6
disciplines (academic institutions) 25, 80, 92, 116–17, 139–40, 150, 212–16, 232
  specialisation 150–1, 214–15
Dostoevsky, F. 55, 248
Dreyfus, Hubert L. 57, 245
Droysen, J. G. 128, 194
Dumit, Joseph 113n.37
Durkheim, Emile 87
  *Les Règles de la méthode sociologique* 86

Ebbinghaus, Hermann 156
'ecology of knowledge' 229n.63
eliminative materialism 108
empirical knowledge 9, 32–3, 36, 38, 48, 62–3, 77, 91, 94, 102, 107, 143, 173, 178, 191, 229, 256
enlightened thought 17, 38–9, 55–6, 86–7, 123, 131, 136, 140, 208
Epstein, Mikhail 73, 75
Eribon, Daniel 56n.69
Ermarth, Michael 42n.44, 80, 124, 152n.47, 155n.54
essential attributes 26–8, 30, 35, 37, 48, 52, 62, 73, 84, 192, 249
ethics, discipline of 145–6, 148
Evernden, Neil 18, 237, 239, 242n.92
evolutionary biology 6–7, 16, 18, 27–35, 61, 90, 107, 194–5, 200–1, 210, 218, 237, 241, 243–4
evolutionary psychology 6, 28–9, 33, 210, 218–20
existentialism 49, 58
expressivist (theory of knowledge) 64, 255

'fallacy of misplaced concreteness' 7
Farber, Paul Lawrence 210n.26
Ferguson, Adam 143
Feuerbach, Ludwig 43–4, 45, 205
Fichte, J. G. 40–1, 49, 138, 246, 253–4
fiction/fact distinction 179, 182
Flanagan, Jr, Owen J. 33, 75
Ford Foundation 212
Ford, Henry 9
Fosdick, Raymond B. 221
Foucault, Michel 7, 120, 142, 173, 240, 245, 248
  *Folie et déraison* 56
  *Les Mots et les choses* 24, 56–8
  philosophical anthropology, opposition to 13, 55–60
  reflexive subject 78–9
Fox, Christopher 3n.4
Frankfurt school 53
Fraser, Mariam 71n.24
freedom, moral and philosophical 38, 40–2, 49, 58, 79–80, 134, 136, 246–8
Freud, Sigmund 185

Gadamer, Hans-Georg 14, 120, 132, 150, 155, 164, 173, 181, 184, 193–4, 222–3, 235, 243, 246, 247n.14, 248, 250
  *ethos* 115–16
  hermeneutic circle 67–8
  metaphor 202–3
  and Mill, J. S. 149
  translation 125–6
  *Wahrheit und Methode* 166–72
Gallie, W. B. 178, 195
Galton, Francis, and nature-nurture 18
Geertz, Clifford 35, 52, 103, 111n.30, 205–6, 233–4, 241, 252–3
Gehlen, Arnold 48
*Geisteswissenschaft* 24, 122–7, 135, 153, 167, 222, 242
gender 10, 21–2, 83–5, 231
Geras, Norman 44n.46
Gergen, Kenneth J. 82n.59

Gergen, Mary M. 82n.59
Giddens, Anthony 75n.33, 88
Goethe, Johann Wolfgang 43, 136, 228
Good, James M. M. 22, 213n.32, 229n.63
Gould, Stephen Jay 91
Gouldner, Alvin W. 82n.58, 88
Grafton, Anthony 132
Gray, Asa 218
Greenfield, Susan 33
Groethuysen, Bernard 243
Gusdorf, Georges 21, 59, 78, 141

Habermas, Jürgen 125, 182, 244
  'interests' 53–5, 226, 230–1
Hacking, Ian, historical ontology 12, 118n.45, 224n.53
Hallberg, Margareta 76n.38
Hampshire, Stuart 65, 72, 77, 89, 100, 120, 175, 246n.12
Haraway, Donna 85n.63
Harré, Rom 101
Hayward, Rhodri 72
Hegel, G. W. F. 5, 41–2, 53, 64–5, 66, 129, 136, 192, 194, 253
Heidegger, Martin 67, 75, 78, 80, 103, 155, 164, 167, 180, 194, 207–8, 230, 233, 238, 240
  and narrative 175, 196
  *Sein und Zeit* 49
Heilbron, Johan 57n.72, 86, 214
Helmholtz, Hermann 149–50, 151
Hennis, Wilhelm 46
Herder, J. G. 40, 64, 122, 131, 134, 136–9, 246, 253–6
  *Ideen zur Philosophie der Geschichte der Menscheit* 136
hermeneutic circle 67–8, 154–5, 171, 183
hermeneutics 67, 132, 135, 137, 151–5, 164, 167–71, 176, 179, 182, 230, 232
Herzen, Alexander 248
Hesse, Mary 97n.6, 113, 226–7, 235, 239
historical ontology 12, 224n.53
historicism 128–9, 194

*History of the Human Sciences* 125
Honneth, Axel 48, 53, 60–1
Horkheimer, Max 241
Howe, Clarence Smith 51n.61
Hull, Clark L. 95
humanism 10, 50, 55–6, 58–61, 75, 120, 232
human sciences
 definition 1–2, 10–11, 13–14, 20, 21–6, 82–3, 114–21, 153, 167, 170–2, 213, 217–18, 222–5, 232, 235, 243, 257–9
 and *Geisteswissenschaft* 122–5, 159–60, 167–72
 history 4, 10–11, 25, 62, 78, 81–3, 141, 170–1, 173–4, 179, 224–5, 232, 249–50, 252
 lack of unity 6, 23
Humboldt, Wilhelm von 137, 152, 154, 184
Hume, David 3, 4, 16, 21, 64
 is/ought distinction 126, 198
 *Treatise of Human Nature* 143–4
Husserl, Edmund 47–8, 52, 153, 157, 164, 169, 238, 240
Huxley, T. H., 'man's place in nature' 16
hypnotism 78

idealist knowledge 40, 42–3, 45, 49, 63, 124–5, 138, 146, 192, 250, 253–6
*idéologues* 141–2
idiographic knowledge 158–9, 164, 234
inductive argument 93, 147, 150
inheritance, human 9, 27–8, 35, 241–2
Institut National 141
instrumental theories 51, 54, 164–5
interests, knowledge constitutive 54, 200, 209, 226
Izenberg, Gerald N. 255n.36

Joas, Hans 48, 53, 60–1
Jung, C. G. 27, 53

Kant, Immanuel 35–40, 42–3, 49, 52, 55, 56, 58, 63, 79, 84, 109, 116, 136, 164, 166, 225, 245, 247, 253
 *Kritik der Urteilskraft* 8, 35–6, 154, 168
 pragmatic anthropology 36, 56
 'What is man?' 8, 35–6, 62, 191
Kaufmann, Walter 49
Kay, Lily E. 221n.47
Kelley, Donald R. 20n.5, 131n.15, 183n.27
Kelly, Aileen M. 248n.16
Kilani, Mondher 252
Kolakowski, Leszek 198
Krohn, Wolfgang 71n.23, 251n.22
Kuhn, Thomas S. 214
Kuper, Adam 9n.12, 30
Küppers, Günther 71n.23, 251n.22
Kusch, Martin 77n.39, 82, 112, 211

Lacan, Jacques 21, 256
Langer, Susanne K. 249
language 2, 30, 57, 107, 115, 170–1, 250
 expressive 137–8
 forms of life 100–1, 110, 201–2
 and gender 84
 of psychology 81–2
 schizophrenic 177–8
 world-constituting 7, 64, 69, 244
Laplacean fallacy 204
Latour, Bruno 21, 70, 118, 119n.48
Lawson, Hilary 69n.17, 75n.36
Leterrier, Sophie-Anne 142n.33
Lévinas, Emmanuel 235
Lewes, G. H. 114
life philosophy 45, 52, 157
life-world 51, 238
Lloyd, Genevieve 85n.62
Locke, John 63, 64, 89
Lotze, Hermann, *Mikrokosmus* 43
Luhmann, Niklas 251n.22

Macaulay, Thomas Babington 21
McCloskey, Donald M. 77n.40, 180n.21, 217n.39, 245
McDonald, Henry 69

MacIntyre, Alasdair 78n.43, 99, 211
  and narrative 175, 184, 188,
    223–4
Magnusson, Lars 57n.72, 86n.65
Malik, Kenan 9n.12, 89n.76
'mandarin' academic culture 52,
  127–8
Mandelbaum, Maurice 193
Manicas, Peter T. 144n.37, 250
'man's place in nature' 16, 34, 42,
  218, 236–7
Margolis, Joseph 26, 76, 194
Markus, Gyorgy 183n.28, 214–15,
  219–20
Marx, Karl 8, 71, 73–4, 136, 205,
  245, 246, 253–5
  and anthropology 43–5, 46, 53,
    191–2
Marxism 48, 56, 70
materialism 107–14
Maturana, Humberto R. 251
Mayr, Ernst 28
Mazlish, Bruce 22, 71
Mead, G. H. 111
meaning 48, 52, 103–7, 113, 152–4,
  166, 169–70, 178, 238, 241–2,
  249
  in stories/histories 5, 191–5,
    203–4, 223–4
Medawar, Peter 97
Megill, Allan 69n.18, 77n.40,
  180n.21, 217n.39, 245
metaphor 200–5
method, scientific 74, 77, 94–5, 97,
  167–8, 215
Midgley, Mary 27, 114n.38
Mill, John Stuart 142, 145–9
  *A System of Logic* 93, 126–7
Millar, James 143
Mink, Louis 181, 184n.30, 188
mirror (theory of knowledge) 63,
  139, 216, 244
modernity 45, 57–9, 86–8, 195–6,
  239–40
Molander, Bengt 76n.38
Monod, Jacques 207
Montaigne, Michel de 89
Montesquieu, baron de 142

moral agency 232–5
moral facts 10, 118, 200
moral philosophy 130, 139–48
moral sciences 126–7, 135, 143–50
Morawski, Jill G. 82n.58, 85,
  231n.68
Mozley, J. R. 238
Müller-Sievers, Helmut 254n.30
Münchhausen, Baron 244
Murdoch, Iris 75, 246
Myers, Greg 180n.21

Nagel, Thomas 67
Napoleon 141
narrative 14, 104, 174–81, 188–91,
  196, 205–6, 223–4
  purposes of 181–5, 197, 224–9,
    235–6
naturalism 9, 11, 95, 98, 107–14
'nature', meanings of word 16–19
nature and nurture 18
Needham, Joseph 236
Nelson, John S. 77n.40, 180n.21,
  217n.39, 245
neo-Kantian philosophy 129
neuroscience 6, 33, 98, 108–10,
  113–14, 210, 256
New Testament, historical
  scholarship 129, 151
Newton, Isaac 95, 135, 204
New World, the 89
Nietzsche, Friedrich 55, 66, 161,
  205, 207–8, 255
nomothetic knowledge 158–9, 164,
  234
Nowotny, Helga 71n.23

Oakes, Guy 124, 158–9, 162n.65, 217
Oakeshott, Michael 178
O'Hear, Anthony 97
Olausson, Lennart 76n.38
organicism 118
Ortega y Gasset, José 49, 50, 192
Orwell, George, *Nineteen Eighty-
  Four* 228

Pagden, Anthony 25
Pascal, Blaise 3, 21

phenomenology 55, 66, 118, 164, 175
*philosophes* 86, 140
philosophical anthropology 13, 35–61, 173, 240, 245, 250, 256
philosophy of history 39, 128, 138, 184, 197
Pickering, Andy 72
Pinch, Trevor J. 70n.20
Pinkard, Terry 254, 255n.34
Pinker, Steven 6, 11, 33, 218, 249
Plato 84, 198, 250
Platonic forms 17
Pleasants, Nigel 230
Plessner, Helmut 46, 48
*poiesis* 179, 251–3, 257, 259
Polanyi, Michael 204, 207, 225, 249
political economy 126, 143–5, 147
Porter, Roy 3n.4
positivism 4–7, 88, 107, 129, 131, 171, 175–6, 226
    Comtean 93–4, 142
    logical positivism 94, 96
postmodernity 69, 71, 73, 232, 248
pragmatism 54–5
presentism 183
Prigogine, Il'ya 251
progress, moral and social 30, 136–7, 148, 185, 218–24
psychoanalysis 6, 21, 185, 244
psychological sciences
    history 24, 80–2, 124–5, 140, 220
    reflexivity in 74–83
    scientific standing 20–1, 76–7, 78, 95, 104, 159
    social psychology 101, 104, 111, 213
Putnam, Hilary 65, 104, 118, 214, 228, 258
    and values 190, 211–12

Queneau, Raymond 189

Rabinow, Paul 57, 245
race 88–91
Randall, Jr, John Hermann 95
realism 12, 17, 65, 76, 117–18, 194

reduction 6, 97, 99, 110, 219, 226
reflection 2–3, 8, 63–8
reflexive consciousness 29, 32–3, 38–9, 41–2
reflexive culture 35, 68–74
reflexivity 8, 13, 57, 62–83, 84, 88, 91, 114–21, 177, 192, 217, 223, 231, 237–8, 248–9, 254, 257
relativism 12, 129, 161, 217
religious faith 7, 10, 11–12, 26–7, 55, 128–9, 158–9, 243–4
    God and man as creators 133–4
    and hermeneutics 151
Richard, Nathalie 183n.27, 216n.38
Richards, Graham 80–1, 91
Richards, Robert J. 43n.45
Rickert, Heinrich 52, 162–4, 166, 209, 242
    *Kulturwissenschaft* 125–6, 159–61
Ricoeur, Paul 104–5, 174n.2, 175, 234
Ringer, Fritz 52, 161n.64
Ritter, Joachim 123n.1
Roberts, Richard H. 22
Rockefeller Foundation 212, 221–2
romantic thought 40, 79, 124, 136, 138–9, 253
Rorty, Richard 55, 63n.5, 208, 238
Rose, Steven 114
Ross, Dorothy 212n.31
Rouse, Joseph 180–1
Rousseau, Jean-Jacques 86–7
Ruse, Michael 30n.24

Samelson, Franz 91n.77
Sandywell, Barry 73, 88n.72, 271
Sartre, Jean-Paul 255, 257–8
Sass, Louis A. 79–80, 178
Scheler, Max 46, 48
Schelling, F. W. J. 138
Schleiermacher, F. D. 151–2
Schnädelbach, Herbert 48, 128n.12, 129n.36, 193n.45
Schneewind, J. B. 139n.28
science, Anglo-American and European conceptions 1, 13–14, 19–21, 42
'science of man' 3, 21, 136

Index 287

*sciences de l'homme* 20–1, 23–4,
135, 140, 142
*sciences humaines* 20, 142
*sciences morales* 140–2
scientific writing 179–81
Searle, John R. 108–9, 117
Secord, P. F. 101
Seeley, John 148
Seigel, Jerrold 62n.1
self, the 3, 5, 41, 58–9, 62n.1, 64,
71–2, 173
self-confirming argument 4, 8, 67
self-creation 8, 14–15, 244–6, 253–5
self-knowledge 1–3, 9, 224–5
self-organising systems 71, 251–2
Shaffer, Elinor R. 150n.46
Shamdasani, Sonu 80n.52
Shapin, Steven 118, 222
Shapiro, Jeremy J. 65–6, 125
Shelley, Percy Bysshe 235, 258
Sherrington, Charles S. 48–9, 200
Singer, Wolf 114
Sirotkina, Irina 248n.16
Skinner, B. F., *Beyond Freedom and
Dignity* 228
Skinner, Quentin 89, 106n.25
Smith, Adam 142–3
Smith, Barbara Herrnstein 67n.12,
217n.39
Smith, Roger 11n.14, 62n.1, 81n.54,
82n.56, 115n.40, 139n.28,
186n.34, 202n.8, 234n.72
social relations 44, 54, 111–12
social rules 29, 98–9
social sciences
philosophy of 29, 92, 97, 98–107,
108, 110, 154, 162, 208–10
scientific standing 94–5, 104
society, as category 85–8
sociology of science and technology
70, 106, 181–2, 207
soul 26–7, 36–7, 40, 53, 136
Soviet Union 26, 96
Staeuble, Irmingard 82n.59
Starobinski, Jean 188n.35, 225,
247–8
Steedman, Carolyn 177n.14
Steinthal, H. 156

Stichweh, Rudolf 212, 214
Still, A. W. 229n.63
Stone, Lawrence, revival of
narrative 174
story-telling 184–91
structuralism 35
Stumpf, Carl 156
symbolic forms 35, 49–50

Taine, Hippolyte 165
Tauber, Alfred I. 12, 232, 235
Taylor, Charles 26, 62n.1, 64, 77–8,
98, 99–100, 120, 137n.24, 175,
198n.1, 253
technological change 61, 69–74, 119
teleological reasoning 39–40, 42–3,
154, 176–7, 254
*telos* 18, 197
'theory' in humanities and social
sciences 7, 238–40
theory-laden knowledge 64–5, 97,
107
Thoreau, Henry David 232–3
Toews, John Edward 42n.44
Tooby, John 28–9
'Psychological foundations of
culture' 9n.12, 218–22
Trigg, Roger 6
Turkle, Sherry 71n.24

Uexküll, Jakob von, and *Umwelt*
51
unconscious, the 3, 7, 53, 78
universities, history 139, 143–4,
148, 150
utilitarians 146

values 14, 129–30, 157–66, 204,
217–18, 225, 241
definitions of 'the good' 29, 32
is/ought 144, 198
life as value 48, 211
and modern science 236–42
moral value of historical
knowledge 229–33
truth as a value 206–12, 216
value-fact relation 97, 198–200
Varela, Francisco J. 251

Vico, Giambattista 8, 122, 172, 173, 192, 200, 238, 240, 244, 246–7
  'new science' 14, 131–6, 193
  *Una scienza nuova* 132
Vienna Circle 94
Virchow, Rudolf, and race 90
Vogt, Karl 149
*Volk* 137
*Völkerpsychologie* 156

Wagner, Peter 88
Wallace, Alfred Russel 34
Wartofsky, Marx 112
Watson, John B., behaviourism 77
Weber, Max 103, 105–6, 199, 226, 239–40, 247
  'disenchantment' 46, 129–30
  and values 161–3, 207–10
Whewell, William 19, 97, 145–8
White, Hayden 176, 179–80, 192–3
Whitehead, A. N. 6–7, 17, 72, 118, 211, 236–8, 240

Whitley, Richard 88n.72
Wilson, E. O. 9n.12, 16
Winch, Donald 21n.9, 148n.42
Winch, Peter 83, 98, 106, 120, 232
  *The Idea of a Social Science* 100–2
Windelband, Wilhelm 127, 158–61, 234
Witt, Charlotte E. 84n.61
Wittgenstein, Ludwig 68, 204, 238
  *Philosophical Investigations* 100
Wittrock, Björn 57n.72, 86, 88n.72
Wokler, Robert 3n.4, 57n.72, 141
Wundt, Wilhelm 76, 156

Yale Institute of Human Relations 212
Yeo, Richard 19n.4, 147n.40
Young, Robert M. 236–7

Zammito, John H. 36n.35, 63n.4, 138n.25